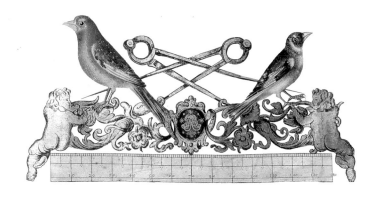

MAGNIFICENT MAPS

MAGNIFICENT MAPS
POWER, PROPAGANDA AND ART

PETER BARBER AND TOM HARPER

THE BRITISH LIBRARY
2010

First published in 2010 by
The British Library
96 Euston Road
London NW1 2DB

On the occasion of the exhibition at the British Library
'Magnificent Maps: Power, Propaganda and Art'
30 April – 19 September 2010

British Library Cataloguing-in-Publication Data
A catalogue record is available from the British Library
ISBN 978 07123 5092 1 (hardback)
ISBN 978 07123 5093 8 (paperback)

Designed and typeset by Andrew Shoolbred
Colour reproductions by Dot Gradations
Printed and bound in Italy by Printer Trento S.r.l.

Photographic acknowledgements

(a=above, b=below, r=right, l=left)

akg-images 18, 146; akg-images/Cameraphoto 48; akg-images/Erich Lessing 13;
Alinari Archives, Florence 82b, 161r; Amsterdam City Archives 101b; © Stichting
Koninklijk Paleis te Amsterdam 21; The Art Archive/Gianni Dagli Orti 27; The
Art Archive/Museo della Civilta Romana Rome/Gianni Dagli Orti 11l; Peter Barber
11r, 160r; bpk/Gemäldegalerie, Staatliche Museen zu Berlin/photo Jörg P. Anders
120; English Heritage Images 121; Matthias Kabel 10; Giovanni Lattanzi 12;
www.leuphana.de/ebskart 80; National Gallery, London 102L, 118; Bibliotheque
Municpal de Lyon. PA 32, folio 1. Photograph Didier Nicole 14; Museo Nacional
del Prado, Madrid 68; Patrimonio Nacional, Madrid 22b, 24; Archivo Storico
Fotografico Soprintendsza BAPSae di Napoli e Provincia 100, 101a; © NTPL/
Andreas von Einsiedel 82a; © NTPL/Mark Fiennes 56–9; © NTPL/Derrick
E. Witty 119; Sandro Rafinelli, www.sandrorafanelli.com 8; RIA Novosti 17;
© RMN/Gérard Blot 15b, 22a; Photo Scala, Florence 20, 26, 49; Alex Segre 160l;
Wikipedia Commons 161l; Zentralbibliothek Zürich Graphische Sammlung 83.
All other images from the collections of the British Library.

Note to reader

A graphic scale-bar accompanies some of the maps to highlight
r size. This is shown by a figure standing by the approximate
portion of the map.

CONTENTS

ACKNOWLEDGEMENTS   7

INTRODUCTION   9

Chapter 1   **THE HISTORICAL BACKGROUND**   11
Classical Antecedents   11
Medieval Europe   13
The Renaissance   14

Chapter 2   **CONTEXTS: THE PALACE**   20

THE GALLERY   26
Henry VIII's map of Italy (Venice, *c.*1425–50)   28
An image to stick in the mind (Venice, 1500)   30
Defence into attack (Nuremberg, 1530)   32
All that glitters… (Venice, 1570)   34
An intimidating map (Paris, 1704)   35
Empire's attributes (Amsterdam, 1617)   36
A Medici beauty (Rome, 1584)   38
World power by association (Venice, 1582)   40
Enlightened colonialism (Amsterdam, 1647)   42
A conquered kingdom (London, 1708)   44
The sport of kings (Hanover, 1717)   46

THE AUDIENCE CHAMBER   48
Impressing and questioning (London and Venice, 1804)   52
A free imperial city (Augsburg, 1521)   54
Lands, friends and loyalty (Mortlake *c.*1663)   56
Spheres of influence (London, 1592)   60
An explanation of power (London, 1603–04)   62
A nation with a map (New Haven, 1784)   64
Pomerania matters! (Amsterdam, 1618)   65
A famous victory (Antwerp, 1627/8)   68
Glorifying the kingdom not the king (Prague, 1722)   70
An impression of power (Milan, *c.*1760)   74

THE BEDCHAMBER AND OTHER 'PRIVATE' ROYAL RECEPTION ROOMS   76
A royal wall map in miniature? (Westminster *c.*1265)   78
A fragment of a royal mappa mundi (Westminster, *c.*1290)   79
Brunswick and the world (1300)   80
New wine in an old bottle (London, 2008)   81

THE CABINET   82
The world for a king (Normandy, 1550)   84
A Renaissance mash-up (Rome, 1551)   86
A Renaissance underpinning (Antwerp, 1564)   88
A souvenir (Paris, 1616)

A reminder of home (Hannover, 1726–7)   91

'A vast book of Mapps' (Amsterdam, 1660)   92

The world in your hand: Globes (1623–1831)   94

Miniaturization: Coins and medals (1580–1971)   96

A free-standing atlas of Europe (London, 1750)   98

Chapter 3   CONTEXTS: BEYOND THE PALACE   100

THE SECRETARY OF STATE'S ROOM   101

Security (London, *c*.1677)   103

Safeguarding the key to England (Dover, 1552)   104

Trying to win support from Mr Secretary Cecil (Nottingham, *c*.1558)   106

Annotated proof (Nuremberg, *c*.1563)   108

Clemency, patriotism and self-promotion (Edinburgh, 1746)   110

The art of war (London, 1777)   112

The finished product (Kolkata, 1785)   114

Cartographic coaxing (St Petersburg, 1869)   116

THE MERCHANT AND LANDOWNER'S HOUSE   118

The world and four continents on a wall (Amsterdam, *c*.1680)   122

The currency of culture (Amsterdam, 1695)   124

The pious businessman's second home (Antwerp, 1570)   126

The European world view (Amsterdam, 1617)   128

A step on the ladder (London, 1724–5)   130

Royal London (London, 1682)   132

The insider's tourist map (London, 2008)   134

The pride of ownership (Bramford, Suffolk, 1582)   138

William Penn's neighbour (Canterbury, 1707)   140

Business concluded (London, 1687–8)   142

Where to send your fishing fleet (Wapping, 1693)   144

For the foreign market (Canton, *c*.1770)   145

THE SCHOOLROOM   146

Teaching piety and patriotism (Evesham, 1390–1415)   148

The Victorians in colour (London, 1851)   149

A princely education (London, 1741)   150

On its last legs (Paris, 1767)   152

A lesson for the teacher (Warsaw, 1903)   156

Cartographic capitulation (Leipzig, 1937)   158

Chapter 4   OUT IN THE OPEN: MAPS FOR THE MASSES   160

A British general election (London, 1880)   162

European peace – on France's terms (Paris, 1869)   163

Russian, Prussian and Churchillian octopuses (1877–1942)   164

Bolsheviks and big business (1922–40)   166

BIBLIOGRAPHY   170

INDEX   174

ACKNOWLEDGEMENTS

This book could not have been written without the support and advice, always generously given, of numerous people. Those named will know why they are mentioned with great gratitude by us: Philip Attwood, Dr Nicholas Baron, Roderick Barron, Ashley Baynton-Williams, Dott. Ermanno Bellucci, Paul Cohen, Dr Catherine Delano Smith, Professor Matthew Edney, Isabelle Egan, Dr Imre Galambros, Antony Griffiths, Martin Halusa, Lesley Hanson, Vanessa Hayward, Francis Herbert, Graham Hutt, Alexander Johnson, Mark McDonald, Professor Marica Milanesi, Sam Mortlock, Felicity Myrone, Martin Oestreicher, Katya Rogatchevskaia, Renae Satterley, Dr David Starkey, Elisabeth Stuart, Professor Günter Schilder, Hilary Turner, Professor Vladimiro Valerio, Dr Ulrike Weiss, Joe Whitlock-Blundell, Thomas Woodcock and Laurence Worms.

We would also like to thank the director and trustees of the British Museum, the direction of the Capitoline Museum in Rome, the Viscount Coke, the College of Arms, London, the Duchy of Cornwall Office, the Masters of the Bench of the Honourable Society of the Middle Temple, the Sheldon Family Trust and the curators and managers at Oxburgh Hall, Norfolk. Our special thanks go to our colleagues in the British Library's Map Library, Susan Dymond, the exhibitions team led by Alan Sterenberg, our colleagues in conservation, the imaging team led by Chris Lee and particularly Catherine Britton, Sally Nicholls and David Way in British Library publishing. We wish to thank Apax Partners and the Patrons and Friends of the British Library for their generous financial support for the exhibition that accompanied this book.

Lastly we would like to thank our families and colleagues for their patience and forbearance while this book was being written and would like to dedicate the book to Professor Juergen Schulz who in the 1970s pioneered a new way of looking at medieval and early modern wall maps.

INTRODUCTION

Many of the most beautiful and technologically advanced maps ever created were intended for display on walls – some were even painted onto walls, or carved onto marble which was placed on walls. Their scientific and aesthetic distinction, however, tended to be in direct contrast to their practical use. Located as they were in galleries, or as ceremonial backdrops, most onlookers had no chance to get close enough to study them in detail. Even when close access was possible, such maps were often so big that only the parts at eye level could be examined in detail. If they were drawn on paper or parchment maps could, of course, be taken down for closer study, but even then there were difficulties caused by their size.

Writing in the late 1820s James Welsh, a former British officer in the East India Company's Bombay army, remembered visiting the surveyor general of Bombay, Charles Reynolds, just when he was working on his first map of India. Welsh recalled that 'I had the gratification of *crawling over a map* fourteen feet long and ten feet broad; to do which, without injury to a production intended to be presented to the Court of Directors [of the East India Company] he [Reynolds] furnished me with silk stockings for hands and feet; and cased in these I moved at pleasure, stopping at particular spots for information'.[1] Despite its size a later version of the map was sent to London for display on the wall of a room 'appropriated for that purpose', though it no longer survives.[2] One may well ask about the justification for producing such handsome maps that could be of no practical use, and wonder why they were so treasured by the people who commissioned them at great cost to themselves, or who received them as gifts. The explanation for the paradox of expense and an almost total lack of utility is that many of these large maps were not primarily intended to provide geographical or locational information, but instead served broader cultural, political and personal purposes.

It should be made clear at the start that some large maps were never intended for public display, even if their individual sheets could be stuck together to create an integrated image. This is almost certainly true of one of the greatest treasures of the British Library, the enormous manuscript map of the mainland of Scotland compiled under the direction of General William Roy between 1747 and 1755. Though the presentation or 'Fair' copy[3] was exquisitely drawn and coloured by the artist Paul Sandby (1731–1809), who has been described as 'the father of English watercolour painting', there is no evidence that it was ever assembled for display in the eighteenth century. Instead it remains for ease of consultation in its original sheets, which are housed in purpose-made boxes.

In this book we shall be examining maps that were intended to be displayed for the most part vertically, and not viewed horizontally for private study as atlas maps are. Display maps include wall maps of all kinds produced on a wide range of materials, from those painted directly onto plaster or on canvas and mounted on walls, maps engraved on marble and placed on walls, tapestry maps and maps painted or printed onto vellum and paper which constitute the most common form of wall maps. Some display maps were to be found engraved on floors and some adorned ceilings. A few were mounted on screens which were intended to be displayed vertically. Most wall maps were large – some very large – but size was not the only qualification for display. Several were quite small, though rarely so small that they could easily be accommodated within the covers of an atlas. The 'great age' of the European display map fell between about 1450 to about 1800, although spectacular examples were created before then and are still being made in the twenty-first century.

Their principal quality was the ability to impress. This was often achieved by their splendour and the artistry that went into their creation. We

The *Sala del Mappamondo*, Palazzo Farnese, Caprarola

The map gallery, Toskanatrakt, Archbishop's Palace (*Residenz*), Salzburg

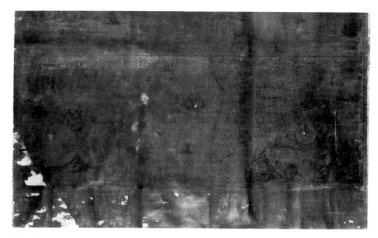

The effects of varnish on a late seventeenth-century Dutch printed wall map.
British Library Maps Roll 529

shall be examining why they have generally been overlooked by scholars and the general public alike, and will be looking at their evolution from earliest recorded historical time (prehistoric wall maps are excluded). We shall then turn to the differing settings for which the various types of display maps were intended. The introductory texts are followed by discussions of specific examples.

Striking and often magnificent though display maps are, their significance has tended to be overlooked because so few have survived. Some of the oldest examples, engraved on marble tablets, have shattered into fragments. Those maps that were drawn or printed on paper were particularly susceptible to mishandling and damage because of their size. Wall maps were usually varnished in the first flush of the owner's enthusiasm to display them to maximum effect, and in the misguided hope that the varnish would protect the image from light and heat. It was not long before the colours clouded over or faded, the varnish darkened and extended exposure to light caused the paper to decay and destroyed the image. Where varnish was not the problem, damp and/or soot from nearby hearths often was. Maps hanging from walls also frequently collapsed under their own weight or, in the case of maps that were kept rolled, got torn as a result of being pulled down from their wooden holders once too often.

Even where maps survived these immediate hazards, perceived obsolescence in style or content took its toll. Many maps were discarded or relegated to the archives and exposure to dust and rats. Most existing wall maps owe their survival to the fact that their sheets were never trimmed, assembled and hung from a wall – even though they were originally and primarily intended to be display objects, adorning a room and often, subtly, conveying favourable messages about their subject-matter, and often their owner also. Instead the individual sheets were kept in boxes or bound into books.

Maps that were painted on walls usually did not survive for many decades. If they were in loggias, exposure to the elements usually did irreparable damage. In interior spaces the maps might be over-painted to update them, or completely painted over – and sometimes even destroyed – when the rooms were put to different uses or divided up. Quite often the maps might be stripped of the decoration that had been intended to give them a context. This is what happened in the late eighteenth century to the map gallery in the Archbishop's Palace in Salzburg, under the influence of the fashionable neoclassical creed of simplicity and the growing assumption that maps were to be valued purely for the accuracy of their geographical depictions. The gallery had been created in 1614 in imitation of the *Terza Loggia* in the Vatican. Deprived of the figurative, botanical and wild-life paintings that had been painted above, below and around the maps, and on the ceiling, as well as on the baroque plasterwork, it has been reduced to a sterile space lined by
 lated and inconveniently placed maps.

Scholarship has not, until recently, served the great wall map well. Aesthetics, connoisseurship and art on the one hand and science, mathematical measurement and maps on the other were long considered entirely distinct and separate. Until a few decades ago, map experts intentionally ignored – or were oblivious to – the broader cultural and political context of all maps. The only standards by which they were prepared to judge a map were its practical use, the quality of its geographical information and the mathematical precision of the underlying survey. The only additional point of interest tended to be the map's publishing history. Detailed descriptions were given of its various editions and the successive states of the copperplate from which the map was printed, with little, if any, consideration of the reasons for the changes. The most reputable map historians encouraged their readers to dismiss out-of-hand as 'mere decoration' some of the most striking features of wall maps.[4] In the rare instances when the great mural maps received detailed attention – for instance from the great Italian map historian Roberto Almagià, who studied the map cycles in the Vatican and elsewhere in Italy in great detail in the 1950s – the focus was purely on their geographical content, their mathematical construction and accuracy and their cartographical sources. Art historians ignored maps entirely since they were considered to be works of science and not art. Since the 1980s the pendulum has finally swung back to some extent and scholars such as Svetlana Alpers, Francesca Fiorani, Peter Meurer, Marica Milanesi, Lucia Nuti, Günter Schilder and Juergen Schulz, some of whom would classify themselves primarily as art historians, have drawn attention to the historical and cultural importance of this hitherto overlooked class of map.

Until recently, however, these large maps have been overshadowed in popular awareness by smaller maps in atlases which have been protected from light and, to some extent, from damage by their bindings. But the minute survival rate of all kinds of wall map in no way diminishes the beauty of the large maps or their importance at the time they were created.

1   James Welsh, *Military Reminiscences* (London, 1830), p.243, quoted by Matthew Edney, 'Bringing India to Hand: Mapping an Empire, Denying Space' in Felicity A. Nussbaum, *The Global Eighteenth Century* (Baltimore & London, Johns Hopkins University Press, 2003), p.71. We are most grateful to my colleague Felicity Myrone for bringing this passage to our attention.
2   Edney, p.69.
3   British Library Maps CC.5.a.441.
4   See for example, R.A. Skelton, *Maps: A Historical Survey of Their Study and Collection* (Chicago, University of Chicago Press, 1971). In a series of three lectures Skelton, perhaps the leading map historian of his time, only mentioned aesthetics once, conceding that a map's 'design and decoration may be of interest to the art historian, to the student of calligraphy and typography, even to the social historian and the psychologist' (p.5). In the late 1940s and early 1950s he and Edward Croft-Murray, then the Keeper of Prints and Drawings, had gone through the British Museum's collections of maps, prints and drawings extracting all the 'useful' items from the 'Drawings' and all the 'beautiful' items from the 'Prints' They did irreparable damage to the integrity of the collections of both departments in the process – particularly after 1973 when, with the foundation of the British Library, Maps and Prints and Drawings became parts of different institutions.

Classical Antecedents

One of the earliest European mural maps intended for display is a painted frieze, created in the ancient Minoan city of Thera on the island of Santorini, which was buried in the ashes of a volcanic eruption in about 1500 BC. The surviving fragments, discovered in Akrotiri in 1971 and now in the National Museum in Athens, show three seaside towns, possibly on the north shores of Crete. They may be commemorative in nature since the pictorial map seems to contain depictions of historical, ritualistic or mythological events. Though little can be firmly deduced about its purpose, the frieze suggests that the emotive power of place led to the creation of wall maps from the earliest, and probably even from prehistoric, times.[5]

Several large wall maps survive from Roman times, though none are complete. They have a more direct relevance to later developments in Europe than their Minoan predecessor since their existence was recorded in literature and later centuries maintained a continuing respect for the achievements of Roman civilization. In about AD 200 an enormous plan of Rome, 13 metres in height by 18 metres in width, was cut into no less than 150 marble slabs. Now known as the *Forma Urbis Romae (or 'outline of the city of Rome')*, the plan was placed on a wall, which amazingly still survives, in a side hall of the Temple of Peace. This had been rebuilt following a fire in AD 192 and seems to have been used as the equivalent of a modern town planning department, housing the official municipal archive of property plans, which were drawn on papyrus.

Since the fifteenth century 1,186 fragments of the plan have been discovered, and they are now housed in the Capitoline Museum in Rome. There are enough to gain a good impression of its overall appearance. The plan had sufficient detail, at a scale of approximately 1:240, for important structures to be identified and named and even for the individual pillars surrounding

temples and public buildings and the steps leading up to them to be visible. Its mathematical precision was such that almost all these features could be shown 'in plan' (that is, a plan-like depiction based on a mathematically accurate survey) so that the general appearance is similar to that of a modern large-scale Ordnance Survey map. The most important buildings, however, were shown in perspective, as is the case today on many tourist city plans, and some features seem to have been depicted not as they actually were but as they appeared in what was judged to have been their heyday. Smaller buildings, however, that constituted the majority of the urban fabric seem to have been depicted schematically, and, although there would have been space, no indications of ownership are given.[6]

The same applies to the second great mural map to survive, also in pieces, from Roman times. The fragments were found in the French town of Orange and form part of what were originally three large plans, which apparently depicted a large part of what is now Provence. The plans, of which there were once probably several more, were created after AD 77 in line with edicts issued on behalf of Emperor Vespasian. They seem originally to have been attached to the inside walls of the local tax office (*tabularium*) situated not far from the enormous, and still largely intact, Roman theatre. Their context, then, was similar to that of the plan of Rome. The original survey was made in connection with the settlement in the area of retired Roman soldiers, and the maps show the rigid division of the land into regular holdings dissected by equally regular roads. Only the outlines of some rivers and islands interfere with the rigid symmetry.[7]

Neither of these maps could have been intended for real administrative use. The marble made it virtually impossible for them to be updated, and on closer analysis the plan of Rome, at least, did not have the level of information that would have been required. The lower edge of the plan of Rome was also exhibited at least two metres above floor level, which would have made it impossible for any visitors to study it in detail. But on entering the hall and seeing the plan just 26 metres from where they were standing, visitors would have been impressed, and possibly even overwhelmed, by the inescapable evidence of the size of Rome and the technological skill of the

Fragments of the *Forma Urbis Romae*
of about AD 200

Fragments of the Orange Cadastre of AD 77

inhabitants who had it.[8] The maps in the tax office in Orange would similarly have symbolized the extent of Roman power, as well as the technical and administrative achievements of the Romans. They would have implicitly suggested that the Roman Empire, and its occupation of Gaul, would be as durable and unchanging as the marble into which the map was incised.

In his immensely influential encyclopaedic book *Historia Naturalis*, the Roman statesman and polymath Pliny the Elder, who was killed in Pompeii during the eruption of Vesuvius in AD 71, repeatedly mentioned another large marble map, about 3 metres high and more than 3 metres wide, of which no trace survives.[9] It is not known whether it was painted or cut into marble. The map showed the whole world, with particular emphasis on the Roman Empire. It was created in the decades after 20 BC, initially under the supervision of Marcus Vipsanius Agrippa – an admiral and the son-in-law of Emperor Augustus – and, after Agrippa's death, of Augustus himself. The finished map was displayed beneath a portico on the Via Lata (now the via del Corso) between the Circus Flaminius and the Theatre of Marcellus in Rome. Some modern scholars have argued that the object commissioned by Agrippa was a written description rather than a map, but it seems that such large maps and map-like views may not have been particularly rare in ancient Rome. There is no reason not to see Agrippa's description of the world as drawn rather than written. The Roman historian Livy (59 BC–AD 17), for example, wrote that the Roman general Tiberius Sempronius Gracchus had set up a 'tabula' of the 'forma' of the island of Sardinia containing a depiction of its capture by the Romans in the temple of Mater Matuta in Rome in the second century BC.[10]

The Romans also enjoyed looking at perspective views of towns. Pliny mentioned a painter, during the reign of Augustus, who, in addition to creating landscape paintings for the walls of Roman villas, had 'first introduced the most attractive fashion of painting the walls of uncovered terraces with pictures of seaside cities … giving a most pleasing effect'.[11] These words were to prove of great importance to developments from the fifteenth century since they appeared to provide the classical sanction that was all-important for Renaissance theorists for the practice of decorating the walls of homes and palaces with maps, and particularly with bird's-eye views of towns.

Town views may not seem, to twenty-first century readers, to be the same as maps, but in classical antiquity they were considered to form part of the same loose group of visual materials. The paintings that Pliny described seemed to fit into the category of chorography, or pictorial depiction of a locality, which the influential geographical theorist Claudius Ptolemy classed as a form of map in his *Geographia*, written about a century after Pliny. As recently as 1997 extensive remains, measuring 10 square metres, of a large fresco showing a townscape dating from the late first century were found adorning the wall of a Roman building near Emperor Nero's luxurious *Domus Aurea*, or Golden House, under the Colle Oppio – a park near the Baths of Trajan in the centre of Rome itself. The discovery has not yet been fully published, but the town, which seems to be a somewhat idealized coastal port, is represented in bird's-eye view: a cartographic technique widely used for town plans about 1,500 years later during the Renaissance. Although it was created shortly after Pliny's death, the town view is presumably the sort of painting to which he was referring.[12]

One of the greatest cartographic monuments of ancient Rome, which survives only in a much later (but apparently accurate) copy, is the so-called

Fresco of a bird's-eye view of a Roman town, Colle Oppio, Rome, from about AD 100

Detail from a mosaic map of Palestine, Madaba, Jordan, from about AD 565

in the church at Madaba near Amman in Jordan. Created in about AD 565, it is calculated originally to have measured no less than 24 by 6 metres and to have covered the entire width of the nave. It could be that it had earlier antecedents. In the 1990s a mosaic 30 metres square, showing Mediterranean islands and dating from about AD 300, was found on the floor of the principal room of a Roman villa in Ammaedara in Tunisia.[14] The purpose of this mosaic seems to have been to show the locations of shrines of Aphrodite (Venus) – possibly in connection with an actual or imagined pilgrimage to them. However, the groupings and depictions of the islands do not correspond to reality, either in location or in shape. It is impossible to say whether the mosaic was a one-off flash in the pan or a clumsily conceived example of a group of cartographic images of which others have yet to be found. It does, however, demonstrate that the use of geographically based images in, and on, the floors of reception rooms has a very long history.

Unlike the Ammaedara mosaic, the Madaba mosaic is, at one level, fairly geographically precise. The Palestinian place names and their locations are derived from the gazetteer of biblical place names compiled in about AD 303 by Eusebius, Bishop of Caesaria, and supplemented by more recent travellers' accounts. Like the Ammaedara mosaic, it seems to be associated with the notion of pilgrimage, actual or allegorical. However, as is the case with many large maps through the ages, it transcends the simply geographical. It depicts with considerable accuracy the physical features of the Holy Land, including deserts, hills, lakes and towns, as well as some of its wildlife, and contains references to the life of its inhabitants. It also emphasizes sites mentioned in the Old and New Testaments, their importance underlined by inscriptions and by pictorial depictions of the more significant cities. The Madaba mosaic presents the image of a Holy Land, perhaps for the first time, and its location in the central area of a church suggests that the main reason why the mosaic map was created was spiritual. It may have been intended as an object for contemplation and a symbol and reminder of the divine, perhaps similar in purpose to the icons which are a more familiar, though somewhat later, aspect of Byzantine art.

Medieval Europe

In the centuries between about AD 300 and 650, a succession of scholars – notably Eusebius, Jerome, Orosius and Isidore, Bishop of Seville – tried to preserve and to Christianize the intellectual heritage of classical antiquity. Their work had an important geographical dimension: most events recounted in the Bible, the tales of ancient Greece and the history of the early church had strong associations with particular places. Simple world maps were sometimes introduced as a way of clarifying and amplifying the religious texts. So important was this spiritual context that the German scholar Brigitte Englisch has even suggested (though her interpretation has been challenged) that the actual proportions of the inherited world map, centred on the Mediterranean which was at the heart of the Greco-Roman world and probably derived from the Agrippa map, were altered.[15] In order to emphasize the importance of the Trinity, for instance, Englisch suggests that the distances between the patriarchates of Constantinople, Carthage, Alexandria and Jerusalem may have been distorted so that all lay along the lines of an equilateral triangle, while other religiously significant cities such as Rome and Babylon became united by circles – the symbol of divine perfection.[16]

Superimposed on this framework were the locations of places famed in classical antiquity and, increasingly, references to the peoples, flora and fauna of the world as originally described in Pliny's *Historia Naturalis*, or *Natural History*. His words had been expanded and Christianized by later writers, with his description of the pelican piercing its breast to feed its

Peutinger Table now owned by the Austrian National Library. This originally showed the whole of the Roman Empire, but in a highly distorted form, as though emphasizing the routes across it. For a long time this led scholars to assume that, despite its ungainly format (it was an enormously long roll), it had originally been intended to be a route map. There has been, and continues to be, a lively debate about it, with widely diverging theories being formulated. Recently, however, Professor Richard Talbert has suggested that it may be a copy of a cartographic frieze that decorated the upper walls of the audience chamber of one of the later Roman emperors, with the personification of Rome (shown as a figure) occupying the space at the centre of the rear wall immediately above the imperial throne. The relatively narrow elongated space available between the ceiling and curtains or frescoes below would have dictated the form of the map, but its purpose (like that of the other examples mentioned above) would have been the glorification of the Roman Empire and, presumably, of the particular Roman emperor for whose palace it was made. Indeed, as Professor Talbert has discovered, another version of what must have been the same map was seen in a former imperial palace near Ravenna by a Renaissance writer.[13]

In the eastern Roman, or Byzantine, empire, ruled from Constantinople (Istanbul) – an empire that survived until 1453, nearly a thousand years after the fall of the western Roman empire – large maps probably continued to be created, at least for a few centuries, but in different media and now with a Christian gloss. The best surviving example is the mosaic of the Holy Land

Jean Germain presenting a mappa mundi to Duke Philip the Good of Burgundy,
(illuminated miniature of about 1450). Bibliothèque Municipal de Lyon. PA 32, fol.1

Large world maps, or mappae mundi, cost a lot in time and money to create. Their drafting could involve much preliminary reading and the seeking out and study of other examples in sometimes distant locations. Scribes and – usually separately – artists had to be commissioned to create the final version and to draw and colour its illustrations. Then there were the costs of transportation if the work was carried out in a fashionable workshop. Finally it would need an elaborately carved and quite possibly gilded frame and careful mounting. The necessary materials and workmanship did not come cheap.[19] Even in what is traditionally, though over-simplistically, regarded as the pious Middle Ages, patrons were only prepared to go to this expense if they considered it to be to their earthly or eternal advantage. Piety could certainly be a strong motivation, but personal ambition might also play its part. A handsome and suitably doctrinally correct map would reflect well on the donor were it intended for a church and the donor was an up-and-coming, ambitious cleric. This may well have been the case with the Hereford Mappa Mundi, which contains a plea to pray for one Richard de Bello, prebendary of Haldingham and Sleaford – possibly a relative of the original donor who himself seems to have been a politically ambitious cleric. A magnificently decorated and intellectually sophisticated mappa mundi would have made an equally strong impression in the palace of a king, prince or nobleman, or in the audience chamber in the town hall of a city republic.

From the late thirteenth century a distinctly different style of map, the sea chart, appeared in the western Mediterranean. Initially they seem to have appeared exclusively in a small format: they were painted onto single vellum or parchment skins or were cut up into smaller maps to produce atlases. By 1350 ostentatious examples, drawn by chartmakers and then embellished by professional scribes and artists in workshops, were being commissioned from wealthy merchants for presentation to north European rulers. Gradually the coastal precision of sea charts combined with the traditional features of the mappae mundi to evolve into a transitional form of world maps. These were meant for public display in churches, but also, and perhaps particularly, in the homes and palaces of the wealthy.[20]

The Renaissance

The fifteenth century saw a dramatic rise in the number, prominence and prestige of secular maps and their makers, especially in Italy. There were many reasons for this.

The first was associated with the Renaissance veneration for classical texts, which encouraged the rediscovery and, most importantly, the appreciation of ancient texts that had been partially or completely lost in medieval Western Europe. In the cartographic context the most notable were the descriptive geographical texts of Pomponius Mela and Strabo and, above all, that of Claudius Ptolemy's *Geographia*. The full Greek text of this work was brought to Italy in the early years of the fifteenth century and speedily, if rather clumsily, translated into Latin. Its existence and a rough idea of its contents had been known in scholarly circles for centuries, but now it could be studied in detail.[21] There were extensive and much appreciated lists of ancient place names, and the text provided the instructions and the co-ordinates necessary for creating mathematically precise world and regional maps in a society that was increasingly prizing the use of logic.

Just as important, though often overlooked, the *Geographia* mentioned local or chorographical maps if only to place world and regional maps, which were the subject of the book, in context. Ptolemy argued that there was a hierarchy of forms in which the world and its parts could be depicted. At the top were regional and world maps, which should be mathematically structured. At the bottom of the hierarchy (and not further discussed or

offspring, for example, becoming a symbol for Christ who sacrificed his life for mankind. Eventually, an appropriately placed pelican came to feature on these world maps as both a natural phenomenon and a spiritual allegory. The maps gradually became visual encyclopaedias of the world, particularly illustrating the history of man and placing his relationship to the Almighty in a geographical context.

Such depictions, however, required space. By about 1100 large world maps were being created to accommodate the ever-increasing amount of information, including many references to recent events, that patrons and viewers now expected to find in them. Usually, but not invariably, they were accompanied by separately bound texts explaining and amplifying their content. As well as being painted onto skin (be it vellum from cows or parchment from sheep), the maps were to be found on cloth (the Latin for which is *mappa*), internal and external walls and even floors, as was the case with the map in the audience chamber of William I's daughter, Adele of Blois.

In many cases the earliest recorded locations of these maps was inside or in the vicinity of churches, though it is now thought unlikely that medieval world maps were actually intended as altarpieces.[17] For instance, the Hereford world map of about 1300, still in Hereford Cathedral, may originally have hung in a transept or side aisle next to the choir, as part of the fixtures that surrounded the shrine of the former Bishop of Hereford, St Thomas Cantelupe, at the time when strenuous – and ultimately successful – efforts were being made to get him canonized.[18] The late fourteenth-century tower of the Dominican church in Ascona, Switzerland, has a (now) very faded and very simple map apparently showing the zones into which the globe was supposed to have been divided. However, it would be wrong to deduce from this that because of their spiritual and intellectual content these maps served purely religious or educational objectives, or that the calm of the church or cathedral was their sole setting.

Comparison between geography and chorography in Peter Apian, *Liber Cosmographicus*, 1524, British Library, c.32.g.15

N.A. Monsiau, *Louis XVI giving instructions to La Perouse, 1785,* painted in 1817. Versailles, Chateaux de Versailles et de Trianon

illustrated in the *Geographia*) there were chorographical or local maps which were or should be pictorial. It was Peter Apian, writing in the early sixteenth century, who made the distinction most clearly in a diagram in his *Liber Cosmographicus*, first published in 1524. In it he compared a world map to the whole of a human head and a chorographical map to a part of the whole, in this case an ear or an eye. By implication (though Ptolemy may not have approved), the notion of a picture came from the late fifteenth century to be extended to bird's-eye views like De' Barbari's depiction of Venice in 1500 or the Seld/Weiditz bird's-eye map of Augsburg of 1521 (see pp.54–5), which required mathematical sophistication in their execution. The 'picture', almost imperceptibly, had come to include the pictorial map.

The Renaissance fascination with ancient learning also led to a new, more critical re-examination of the writings of well-known classical authors, which were made widely available after 1450 through printed editions. Euclid's mathematical work the *Elements* enjoyed a dramatic increase in popularity, though it had been known in the West for centuries, with numerous new editions appearing from the 1480s. Ambitious mapmakers came to boast of their familiarity with Euclid's teachings, though few seem actually to have followed them. In the same way, a fresh look was taken at Pliny's *Natural History*, with its mention of the large maps and of the views created for temples and houses in ancient Rome. The latter were now interpreted as being choro-

graphical works as defined by Ptolemy. Pictorial they might be, but they were as much maps as the more mathematically constructed regional and world maps. Indeed, from the fifteenth century onwards world and regional maps, in disregard of the austerity of Ptolemy's theories, frequently contained prominent depictions of peoples and places derived from the descriptive geographical writings of Strabo, Pomponius Mela and Pliny. It was the combined influence of Ptolemy and the descriptive geographers that determined the character and appearance of the grandest manuscript, painted and printed maps that were created in Europe between 1450 and about 1780.

As a result, mapmakers came to be regarded as painters and enjoyed the increased status attached to artists. By the early sixteenth century influential theorists such as Baldassare Castiglione, Macchiavelli and, in England, Thomas Elyot were praising what we would call mapmaking, but they called painting, as a courtly art that was appropriate to a courtier or future governor. Mapmaking itself they described as an indispensable aid for the administrator and general, as well as being a source of pleasure and satisfaction for the prince himself.[22]

A second factor in raising the profile of maps was political and more widely cultural. The competition for resources between city states in the central and north Italian heartlands of the early Renaissance led to a precocious awareness, compared to the rest of Europe, of the practical utility of maps for military and strategic purposes, as well as civilian uses such as the planning of forts, canals and aqueducts.[23] From about 1290 detailed local maps were being created both in the old city republics, notably Venice, and in such cities as Milan, Ferrara and Mantua. These had become dominated by leading families (such as the Visconti, the Estes and the Gonzaga) or by local despots and mercenaries or *condottieri*, such as Francesco Sforza who succeeded the Visconti in Milan. The *condottieri* and heads of the urban dynasties had had experience of the utility of maps and were often well-versed in classical literature. They were also intent on self-glorification as a means to consolidate their disputed holds over power. They came to see wall maps and town views and plans – be they manuscript, printed or painted on their wall, displayed in splendidly decorated apartments lined with figurative, devotional and historic paintings – as one of the most effective media for 'soft' propaganda. At one level the maps conveyed messages of military knowledge and political power and authority over the places depicted. At the same time they reflected intellectual sophistication, aesthetic sensitivity and hence personal suitability for high office. With the rise of dynastic states beyond the Alps in the course of the fifteenth and sixteenth centuries, the same outlook became prevalent among the ruling classes in the rest of Western Europe.

The Great Discoveries were a third factor that contributed to the increased popularity of maps. From 1415 the voyages of Portuguese fleets along the coasts of Africa – in search of a sea route to the riches of the Indies and of heathens to christianize – aroused great interest and considerable envy throughout Europe. This increased still further following the discovery of the Americas. The resulting appetite for information could be to a great extent satisfied by up-to-date maps, thus instigating a rivalry among the ruling elites in Europe to demonstrate their ownership – if not of the lands themselves and their treasures, then at least of the latest knowledge about them.

Finally there was the coming of printing. Many of the most distinguished and expensive display maps continued to be hand-made but a growing number from the late fifteenth century were printed on more than one sheet. The grandest of these multi-sheet printed wall maps could be as decorative as the most splendid manuscript wall map. From the sparse evidence available, they seem initially to have been extremely expensive and available only to the powerful and wealthy. Their regular appearance in descriptions of sixteenth-century palaces suggests that at that period they were almost as esteemed by their owners as manuscript wall maps. They

King Philip II of Spain as Poseidon on Jacopo Gastaldi's world map, *c*.1561.
British Library Maps C.18.n.1

made particularly effective propaganda tools because of the relative ease with which they could be replicated without any changes of content.

The result was a variety of maps that, in terms of content, appearance and purpose, had never been seen before in Europe and which had important repercussions for display maps. At one extreme there were the austere, non-pictorial, mathematically constructed world and regional maps created by Ptolemy; an increasing number of updated (commonly termed 'new) Ptole-maic maps created in the same style; books containing maps and the navigational charts, and administrative and military maps that were intended for everyday use. At the other extreme were large, resplendent display maps which reflected the medieval assumption about the encyclopaedic function of maps. Essentially traditional, spiritual mappae mundi continued to be cre-ated until at least 1540. This view of the nature of maps justified the continued depiction on them of historical and biblical events, as well as of the peoples and wildlife of the world. Such a view also meant that, like their medieval predecessors, the cartographical image frequently conveyed implicit but important theological and philosophical messages. Now, however, the maps included depictions of the natives of the newly discovered lands, some-times based on travellers' accounts, sometimes on prototypes found on

medieval world maps.[24] They were plotted onto coastal outlines that were dependent on observation and the use of scientific instruments, and which gave the latest information about the Western discovery of the world.

The technical expertise, artistic skill and the luxurious materials – including vellum, semi-precious stones, silk, silver, gold, copper and precious woods for the frames – which were needed to create the finest manuscript wall maps did not come cheap. The settings for which they were intended were, accordingly, usually sumptuous, and would normally have been acces-sible only to a select elite whom it was hoped would be duly appreciative – and impressed. Sources for the period 1450–1800, such as private and diplo-matic correspondence, inventories, travellers' accounts, treatises and the rare surviving painted galleries, document the presence of maps in the palaces of rulers and nobility. The same sources, supplemented by newspaper adver-tisements and paintings – particularly Dutch genre paintings of the seventeenth century – record the presence of usually less spectacular printed wall maps in the homes of wealthy landowners and merchants. To be luxuri-ous, the maps did not necessarily have to be large; in many cases small maps, enhanced by flamboyant colour and expensive frames, were to be found in these spaces. It also can not be emphasized enough that maps, whatever their

St George's Hall, The Hermitage, St Petersburg, in about 1967

Charlie Chaplin performing a 'ballet dance' with a balloon globe in *The Great Dictator* (United Artists, 1940)

size, were only one element, and sometimes quite a minor one, in the overall decoration of a gallery, chamber or room.

Handsome wall maps were, however, not only to be seen in such luxurious surroundings. Their creation was sometimes regarded as a career opportunity by their makers, particularly if the person commissioning them was in a position to offer further work or other favours. In other instances it was taken for granted that a king, prince, cardinal or leading minister would expect even working documents to be attractively presented. This helps to explain the existence of maps that, though ostensibly intended to serve very practical administrative, political, diplomatic or military purposes, are finely drawn, beautifully coloured and often embellished with decorations of a high order. These were used behind the scenes in offices and working rooms where

they might decorate the walls, but could also be taken down and examined on a table. A reconstruction by the artist Nicolas-André Monsiau of the scene showing Louis XVI of France giving instructions to the navigator Comte de Lapérouse in Versailles, and now on the walls of the museum there, shows a large and elegant map being used in this way.

There were still humbler contexts for display maps. From at least the fifteenth century, and probably earlier, large maps, often with a simplified content that could be seen from a distance, have been used as instructional aids in school classrooms. While some of these maps were up-to-date and highly sophisticated in construction and content, the more unscrupulous publisher saw schoolroom use as a means of artificially extending the life of old copperplates engraved with out-of-date and inaccurate maps.

Since at least Roman times, the humblest, most populist and arguably the most effective context for cartographic display has been the open air. Some cartographic symbols – and notably globes – have been included as architectural features in public buildings, and the nineteenth and twentieth centuries saw the democratization of large maps. They began to appear as integral features in forceful commercial and political posters, often created for street display by skilled designers aware of the emotive force of cartographic images.

Whether in palaces, homes, offices or schoolrooms, or indeed open spaces, however, the description of the maps in surviving documents is generally brief. It is therefore usually difficult to be certain whether the map mentioned is big or small, manuscript or printed, plain or highly decorated, or framed or unframed unless it is specifically described. Nevertheless, in some inventories an indication of financial value is given, and in several instances there is sufficient information for a long-vanished map to be identified with a fair degree of certainty from modern bibliographies.

The messages conveyed by the maps worked at various levels on the observer. Quite often the 'message' appeared to be purely intellectual, philosophical or scientific. This was particularly the case with maps in the cabinets of palaces or in schoolrooms, though often audience chambers seem to have contained up-to-date world maps. Yet these very maps also served propaganda purposes. These could be overt. Jacopo Gastaldi's world map of about 1561, for example, shows Philip II of Spain in the barely concealed guise of Poseidon ruling the waves (see page 16). Usually the propaganda message was more subtle. A world map may have suggested that the owner was intellectually gifted and well informed about the latest discoveries and the latest surveying techniques. The assemblage of maps and objects in a cabinet of curiosities might similarly have given an impression of the owner's intellectual awareness and even his control of nature through his possession of specimens and knowledge of where they came from. A finely executed map of his dominions might testify not only to his employment of the best available surveyors, but also to the extent of his power and control. In 1591 a visitor to Theobalds – the country mansion of Lord Burghley, Queen Elizabeth I's first minister – noted that the gallery contained a great wall map 'of the Kingdom, with all its cities, towns and villages mountains and rivers; as also the armorial bearings and domains of every esquire, lord, knight and noble who possesses lands and retainers to whatever extent'.[25]

The great age of the European display maps may have passed, but the mentalities, ambitions and vanities that led to their creation remain very much alive. The use of maps to embody the state and to inspire awe has continued into modern times. As recently as 1990 a great map of the USSR, made of semi-precious stones, was displayed in St George's Hall in the Hermitage in St Petersburg, once the principal throne rooms of the tsars (see page 17). The map had originally been created for the 1937 world fair in Paris to demonstrate the rapid industrialization that had taken place in the Soviet Union since the revolution of 1917.[26] In such a location, indeed, the map took the place of the monarch as a symbol of the wealth and power of the state. Following the fall of the Soviet Union, the map was removed from St George's Hall and replaced once again by a platform surmounted by a throne.

Large maps have also played their part in film. They were used to create an aura of power in the opening sequences of *Dr Strangelove* and *Godfather III*, where the American cabinet and mafia gangsters respectively meet with a backdrop dominated by world maps. Great globes, so brilliantly satirized by Charlie Chaplin in his 1940 film *The Great Dictator*, also have

their place. Outsize examples were commissioned as Christmas gifts from the United States Army for President Roosevelt, Stalin and Churchill in 1942.[27]

Since 2000, monumental map-like artworks, inspired to a large extent by the great medieval world maps or reflecting pride and knowledge of London, have been created by such artists such as Grayson Perry (page 81) and Stephen Walter (pp.134–5). Like the paintings by Lucian Freud, displayed alongside old masters at the Duke of Devonshire's palatial home at Chatsworth in Derbyshire, they too can be hung next to the earlier maps which inspired them without appearing out of place.

5   Germaine Aujac, 'The foundations of theoretical cartography in archaic and classical Greece' in J.B. Harley and David Woodward (eds), *The History of Cartography volume one. Cartography in Prehistoric, Ancient and Medieval Europe and the Mediterranean* (Chicago, University of Chicago Press, 1987), p.132. Illustrated: plate 3.

6   See the Stanford Digital Forma Urbis Romae Project on http://formaurbis.stanford.edu/ for details, discussion and reproductions.

7   See http://orange.archeo-rome.com/ and http://www.uea.ac.uk/~jwmp/ceze.html (a computer-aided reconstruction of the landscape). See also, though it is outdated, O.A.W. Dilke, *Greek and Roman Maps* (London, Thames & Hudson, 1985), pp.108–10.

8   See Richard Talbert's essay in *Ancient Perspectives: Maps and Their Place in Mesopotamia, Egypt, Greece, and Rome* [16th Kenneth Nebenzahl Jr. lecture in the History of Cartography] Richard Talbert(ed.) (Chicago, University of Chicago Press, forthcoming).

9   E.g. Pliny *Natural History*, book 3.17 (Loeb edition. Cambridge, Mass., Harvard, 2003). And see O.A.W. Dilke, *Greek and Roman Maps*, pp.41–53 for Pliny's other references to the map. There has been an extensive debate since Dilke wrote on the nature of Agrippa's survey as displayed in Rome, with the general opinion now inclining towards the assumption that it was a map.

10   Juergen Schulz, 'La veduta di Venezia de Jacopo de' Barbari' in *La cartografia tra scienza e arte. Carte e cartografi nel Rinascimento italiano* (Modena, Panini, 2006, revised edition of Italian translation originally published in 1990), p.75 n.67.

11   Pliny, *Natural History*, book 35 (Loeb edition. Cambridge, Mass., Harvard, 2003), pp.116–7 .

12   Eugenio La Roca, 'The Newly Discovered Fresco from Trajan's Baths, Rome', *Imago Mundi* 53 (2001), pp.68–79.

13   See Talbert, forthcoming.

14   Fathi Bejaqui, 'Îles et villes de la Méditerranée sur une mosaïque d'*Ammaedara* (Haïdra, Tnisie)', *Académie des Inscriptions & Belles-Lettres. Comptes Rendus des Séances de l'Année 1997 juillet–octobre* (Paris, 1998), pp.825–60. I am most grateful to Professor Richard Talbert for bringing this to my attention.

15   Brigitte Englisch, *Ordo Orbis Terrae. Die Weltsicht in den Mappae Mundi des frühen und hohen Mittelalters* (Berlin, Akademie Verlag, 2002).

16   More geographically realistic maps continued to be created side-by-side with these encyclopedic maps, but very few survive and there may never have been many.

17   Marcia Kupfer, 'Medieval world maps: embedded images, interpretive frames', *Word and Image* 10/3 (July–September 1994), pp.262–88.

18   See most recently the essays by Martin Bailey, M.B. Parkes and Nigel Morgan in P.D.A. Harvey, *The Hereford World Map. Medieval world maps and their context* (London, British Library, 2006) and Dan Terkla, 'The Original Placement of the Hereford Mappa Mundi', *Imago Mundi* 56 (2004), pp.131–51.

19   See, for instance, Rodney Thomson, 'Medieval Maps at Merton College Oxford', *Imago Mundi* 61 (2009), pp.84–90 particularly p.88 (entries for 1460–1).

20   R.A. Skelton, 'A Contract for World Maps at Barcelona, 1399–1400', *Imago Mundi* 20 (1968), pp.107–113; Evelyn Edson, *The World Map 1300–1492. The Persistence of Tradition and Transformation* (Baltimore, Johns Hopkins University Press, 2007).

21   Patrick Gautier Dalché, 'The Reception of Ptolemy's *Geographia* (end of the fourteenth century to the beginning of the sixteenth century)' in David Woodward (ed.), *History of Cartography volume three. Cartography in the European Renaissance* (Chicago, University of Chicago Press, 2007), pp.285–364.

22   Peter Barber, 'England I: Pageantry, Defense and Government: Maps at Court to 1550' in David Buisseret (ed.), *Monarchs, Ministers and Maps. The Emergence of Cartography as a Tool of Government in Early Modern Europe* (Chicago, University of Chicago Press, 1992), pp.30–2.

23   For a general introduction see David Buisseret, *The Mapmakers' Quest: Depicting New Worlds in Renaissance Europe* (Oxford, Oxford University Press, 2003).

24   Surekha Davies, 'Representations of Amerindians on European Maps and the Construction of Ethnographic Knowledge' (Ph.D. thesis, Birkbeck, University of London, 2008). Summary in *Imago Mundi* 61 (2009), pp.126–7.

25   Frederick, Duke of Wurttemberg, in *England as Seen by Foreigners*, William Brenchley Rye (ed.) (London, 1865), p.44.

26   Dominique Moran, 'Soviet cartography set in stone: the 'Map of Industrialization'. *Environment and Planning D: Society and Space* 24 (2006), pp.671–89. I am very grateful to Dr Nicholas Baron for alerting me to this article and to the existence of this map.

27   Arthur H. Robinson, 'The President's Globe', *Imago Mundi* 49 (1997), pp.143–52. Churchill's example can still be seen as his home in Chartwell in Kent. The globe measures 50 inches (125 cm) in diameter.

CHAPTER 2 CONTEXTS: THE PALACE

The most splendid of the maps intended for display were to be found in palaces not only of royalty, their representatives and the aristocracy, but also those of city states, provincial governments and governors and in town halls. It is extraordinarily difficult to pin down precise locations for most maps as inventories tended simply to list them by palace. Even the mention of a specific room can be of little help since a visitor might refer to the room loosely as an 'audience chamber', where the further description makes it clear that he or she is actually referring to a formal dining room that was used for different purposes at other times.[28] Where the room is specifically named it can be difficult, if the building or the room no longer survives, to locate and understand its function at the time when a map was hung there, since the names and functions of most rooms tended to change over time. If the room is still in existence and its maps are prominently displayed, the name may be tautological. The palace of Cardinal Alessandro Farnese at Caprarola near Rome, for instance, has a room filled with mural maps called the 'Sala del Mappamondo', as do the Palazzo Pubblico of Siena and the Doge's Palace in Venice, though in these cases the maps no longer survive. In other cases the name itself may be unhelpful or misleading. Names such as 'presence chamber' or 'privy chamber' (to be explained in the next paragraphs) continued to be attached to the same rooms at Whitehall Palace for centuries, but their nature changed, as what had once been genuinely private – such as the Privy Chamber – became increasingly public and a suite of more private rooms was created beyond.[29]

Luckily locations can sometimes be pinned down, if occasionally tentatively, from mentions of maps and their settings to be found in such diverse sources as diaries, correspondence (for instance that of the Gonzaga family around 1500),[30] building accounts and the rare mention in books such as Richard Hakluyt's *Divers voyages touching the discouerie of the Americas* (1582), in which he describes some of the maps then on display in the Privy Gallery at Whitehall. From these sources it is apparent that, with the exception of the audience chambers of city states, provincial assemblies and town halls, maps were most usually displayed in private or semi-private apartments.

These can be broadly classified by function. Certain sorts of maps were displayed in galleries that connected different parts of private apartments or linked private spaces (such as bedrooms) and semi-private spaces (such as chapels or private audience chambers). Other types of map were to be seen in what were in effect audience chambers, though they might (particularly in the Middle Ages) have also functioned as bedchambers and later – under the name 'presence chambers' – as official dining chambers. To this group 'privy chambers' also belong. Lastly there were 'cabinets', which could also be libraries or 'wardrobes' or studies, but which represented the most private parts of the official apartments. It was in these spaces that the ruler kept his most prized and prestigious possessions, and to which only his or her most intimate and trusted servants or friends were permitted. The maps to be seen here would also have been particularly special. To some extent the maps displayed in these rooms were interchangeable, but each room seems to

LEFT: The *Galleria delle Carte Geografiche*, The Vatican

RIGHT: The *Burgerzaal*, Royal Palace (formerly the Town Hall), Amsterdam

have had some types of map which were specific to it, while the balance of the interchangeable maps was different in each space.

There was a fourth space containing maps which was originally to be found in a palace, and which in some absolute monarchies continued to be located in palaces until the late eighteenth century. This was the room occupied in earlier periods – and at first only spasmodically – by the royal council or parliament, and later by secretaries of state or ministers or the committees

of government which they chaired. The maps to be found in these spaces had similarities to those to be found in the offices of trading companies, and in many countries the machinery of government and administration had moved out of the palace by 1700, though remaining in close vicinity to it.

The types of map to be found in palaces were many and varied. They varied in size also, and while most could not easily have been bound into books, there seem to have been many that were quite small – no larger than the panel pictures next to which they found themselves hanging. The most spectacular maps were undoubtedly frescoes, adorning courtyards, galleries and cabinets. Other painted maps were to be found framed on panels or painted on canvas, such as the huge map of the surroundings of Bruges executed in oil by Pieter Pourbus, made for the walls of the governor's palace.[31]

There were numerous map tapestries, though very few survive. Created from silk, gold and silver threads and involving much work in their preparation, tapestries were the most expensive decorative artefacts to be found in a palace (and very useful too as draft excluders). Related to tapestry but much smaller was the map of 'Parma with its territory, in needlework, on silk', mentioned as being in Whitehall Palace in 1610, and the 'longe needleworke Qushion of silke of the Cittie of Antwerpe', which was in Hatfield House in 1611.[32] Perhaps both had been gifts to James VI of Scotland from its duke, Alessandro Farnese, when he was serving as governor of the Spanish Netherlands in the 1580s. Where the money could not stretch to tapestries 'painted cloths' were created to perform a similar function, and numerous examples are recorded in the collections of Henry VIII. Several rulers had three-dimensional relief models on display: Henry VIII had a representation of Dover in clay, for example, while George III's models of the royal dockyards of England, once on display in his library at what is now Buckingham Palace, are now to be seen in the National Maritime Museum.[33] Occasionally maps were engraved into marble and placed on floors, as in the reception hall, or *Burgerzaal*, of the Town Hall, now the Royal Palace, in Amsterdam, possibly in conscious imitation of classical precedents.

Most maps came in formats and materials that are more familiar to people today. Globes of various sizes featured prominently, with particularly large examples in the grander royal libraries[34] and reception rooms, and miniature examples in the cabinet. There were plentiful coloured multi-sheet maps on vellum, usually manuscript but sometimes printed, which were reserved for the grandest locations. Although manuscript maps seem gradually to have come to be regarded as the more prestigious for display purposes – at least in a royal context – prior to 1600 finely printed maps on paper were also prized, particularly if they were large. Many can be identified in the inventory of Henry VIII's goods, such as De' Barbari's great plan of Venice of 1500 (see pp.30–1), Francesco Rosselli's earlier multi-sheet map of Florence – of which only an impression from a single sheet now survives[35] – and the Seld/Weiditz map of Augsburg of 1521 (p.54–5).

Where the printed examples were not available 'painted cloths' seem to have substituted for them. Maps were often framed even, it would seem, when they were enormous, such as the 'large Mappe of the whole worlde of parchement sette in a frame of woode havinge the kynges armes therin' that was owned by Henry VIII.[36] Often, however, the maps seem to have hung from the walls or, to guess from their format, were placed on tables to be studied and admired, like Pierre Desceliers' great world map of 1550 (p.84–5), the lettering and images of which are oriented so that they are the right way up when read from either side.

The arrangement of maps in palaces ranged from the highly organized – where the integrated combination of maps, figurative paintings, plasterwork and perhaps statues conveyed limited and specific messages – to the utterly miscellaneous, where paintings and maps were thrown together with no apparent underlying logic. More usually galleries fell between these two

The *Galeries des Cerfs*, Fontainebleau

extremes, particularly north of the Alps where conditions were generally too damp for plaster to dry and it was usually impossible to achieve the tight integration of painted geographical and figurative images and stucco sculpture to be found in Italy. In these northern lands wall maps had to be attached to, or hung from, walls side-by-side with tapestries, figurative, landscape and historical paintings on wood panels or canvas, and sculpture. Though the arrangement could not be as integrated, or permanent, as in the painted galleries of Italy, an effort was nevertheless sometimes made to select and arrange the artifacts saccording to themes which conveyed specific, flattering messages. Thus Henry VIII's maps and map-like paintings in Whitehall Palace, as recorded in his inventories of the 1540s, were used, together with paintings and figurative tapestries, to record his past triumphs, to emphasize his piety and learning, to display his familiarity with his fellow rulers and their homes, to project his patriotism and to demonstrate his up-to-date knowledge of his country and of the world.[37]

There were many instances where a room had just a single map among other non-cartographic *objets d'art* all related to a particular intellectual, religious or political theme.[38] A single mappa mundi in the gallery between the king's bedroom and the royal chapel at Hampton Court in the 1540s, for instance, was surrounded by paintings on religious themes. The whole conveyed a message of conventional royal piety.

... de Pannemaker after Jan Cornelisz Vermeyen, tapestry map showing the western Mediterranean viewed from the North, Royal Palace, Madrid

Philip Apian, *Chorographia Bavariae*
and detail, British Library Maps, K Top. 96.2.2

Rooms with walls covered with fresco paintings provide the best examples of the use of maps as vehicles for propaganda. In these spaces the maps were integrated with the surrounding figurative paintings and stucco work in one impressive, intellectually coherent whole which transmitted an often complex ideological message. The outstanding examples are the *Galleria delle Carte Geografiche* in the Vatican, the *Terza Loggia*, also in the Vatican, the *Sala del Mappamondo* or *della Cosmografia* at the Farnese palace in Caprarola and the *Guardaroba Nuova*, also known as the *Sala* or *Stanza delle Mappe Geografiche*, in the Palazzo Vecchio in Florence. Created between 1560 and 1584, they respectively served as galleries, a reception room and as a rather large cabinet of curiosities.[39] The *Terza Loggia* and the slightly later *Galleria delle Carte Geografiche* deliver clear spiritual and political messages illustrating, respectively, the spread of Christianity throughout the whole world and Italy's special role as a second Holy Land. The *Sala del Mappamondo*, created for Cardinal Alessandro Farnese in his summer palace of Caprarola outside Rome, proclaims Italy to be a second Holy Land by placing the map of Italy next to that of Palestine. The *Sala*'s astronomical ceiling, which looks down on maps of the world, the continents, Italy and Palestine, as well as portraits of the great explorers from Marco Polo to Magellan, more implicitly delivers the Counter-Reformation message that modern science and the great discoveries as encapsulated in maps are

manifestations of divine benevolence. The message of the *Guardaroba Nuova* as originally planned[40] is secular, but no less far-reaching. The maps of all the regions of the world, painted on the front of cupboards, were supposed to be grouped, according to the sequence stipulated by Ptolemy, around a great globe and an armillary sphere, originally intended to descend from the ceiling. Above each cupboard it was planned to display busts of people from the region depicted in the map on the front, with some of the relevant flora and fauna featured beneath. Taken as a whole, the *Guardaroba* was to proclaim the glories of the cosmos.

The imagery was also intended to seduce visitors into drawing parallels between the cosmos and the universal knowledge of its near namesake, Cosimo I de' Medici, Grand Duke of Tuscany, who had commissioned the decoration. Almost invariably these sequences of painted maps delivered more personal, immediate and secular messages, side by side with the grand intellectual or religious ones. The decoration of the *Sala del Mappamondo* in Caprarola, including the astronomical map on the ceiling, is full of personal references to Cardinal Alessandro Farnese which transform it into a visual endorsement of his suitability to become pope (the room was painted shortly after he had failed to win election in 1572). The *Terza Loggia* and the *Galleria delle Carte Geografiche* in the Vatican in their turn proclaim the papacy's spiritual supremacy over Italy and the world, and its role as a

Bernard van Orley, tapestry of 'Earth under the protection of Jupiter and Juno'
from *The Spheres*, Royal Palace, Madrid

bastion against heresy and the infidel – both of which had political undertones.[41] The *Galeries des Cerfs* in Fontainebleau (one created in the 1550s by Henri II but no longer surviving, and the other created by Henri IV some forty years later and still in existence, though heavily retouched) had maps of the principal hunting estates owned by the kings of France lining their inside walls. As well as being suitable decoration for the room from which the kings departed for the hunt, the maps also implicitly reminded the courtiers who accompanied the monarch of his virility, and of his power and wealth as the greatest landowner in the kingdom. It was a particularly pertinent message in a country threatened by powerful magnates that was about to slip into, or was emerging from, civil war.

Such integrated messages were sometimes found in rooms where the maps did not come in painted form. One such was a chamber in a palace occupied by Cardinal Charles de Bourbon in the mid 1580s, just before he was formally recognized by French Catholics as 'Charles X' of France in opposition to the legitimate but Protestant heir, Henri (IV) of Navarre. Between 1575 and 1585 Cardinal de Bourbon commissioned a now-lost tapestry map of Paris, ultimately derived from one painted some fifty years earlier for Francis I. An inventory of the 1640s suggests that it was accompanied by other tapestries containing maps of Venice, Constantinople and Jerusalem, as well as a large tapestry map of Italy.[42] The tapestries could well have been accompanied by allegorical paintings on canvas to create an impression of splendour every bit as great as that of the Italian galleries – but with a medium, the tapestry, that was both more obviously luxurious and appropriate to the colder

climate than a fresco. At a superficial level the map tapestries would have been seen as a means of emphasizing the cardinal's status as a man of learning and a loyal Catholic (through the map of Italy) and also the importance of Paris by comparing it to great cities of the past. More delicately, however, they would have emphasized the cardinal's credentials as a future Catholic king, in much the same way that for most of the past century the Bourbon claimants to the throne of France have linked their names and fortunes to the nation's capital by assuming the title of Count of Paris.

More often the role of maps in a palace was to glorify the ruler in simple and fairly obvious ways. This was often done by highlighting his martial virtues and past achievements through depictions of his triumphs. Many were executed as bird's-eye views seen from an artificially high perspective. These were particularly common in printed and painted form in the early and mid-sixteenth century. The best-known examples are perhaps the quasi-cartographic paintings of the Field of the Cloth of Gold (1521), commissioned by Henry VIII in the 1540s and now in Hampton Court. Tapestries were also enrolled in the service of royal glorification, perhaps the most magnificent being the series woven between 1548 and 1554 by Willem de Pannemaker in Brussels after paintings created between 1546 and 1550 by Jan Cornelisz Vermeyen, commemorating Charles V's campaign in Tunis of 1535.[43] Their successors of the seventeenth and early eighteenth centuries were numerous series of tapestries glorifying the victories of Danish and Swedish kings and of princes and generals such as Charles, Duke of Lorraine, Max Emanuel, Elector of Bavaria or John, Duke of Marlborough (still in Blenheim Palace, but with a related and lesser-known series of paintings in Marlborough House in London). These series show strong cartographic influences in the depiction of their receding backgrounds.

Another descendant of this type were the large and flamboyant manuscript and printed wall maps commemorating specific victories. They continued to adorn palace walls into the eighteenth century and retained strong pictorial elements. A comparison between Callot's printed map of the Siege of Breda and Velazquez's painting of the surrender of Breda, now in the Prado in Madrid, with their prominent figurative foregrounds and cartographic or quasi-cartographic (bird's-eye view) backgrounds, reveals how close painting and chorographic cartography still were to each other in the mid-seventeenth century.

A ruler might also promote himself by displaying maps of his lands and those of his allies. The walls of what is now yet another *Sala delle Carte Geografiche*, but was originally the *Terrazza delle Matematiche* in the Uffizi in Florence, dating from the late 1580s, feature maps painted by Leonardo Buti after designs by Stefano Bonsignori. They portray not only the ancestral, Florentine lands but also the Sienese dominions that Grand Duke Ferdinando's father, Cosimo I, had conquered – and which had led to the Emperor bestowing on him the title of Grand Duke of Tuscany. The impact must have been particularly impressive when the loggia was in its original form and the two remaining sides of the room consisted of an open arcade with a view over Ferdinando's capital, Florence.

The propaganda effect was all the greater if the maps were the fruit of fresh and accurate survey. In the course of the later sixteenth and early seventeenth centuries, rulers often commissioned the first detailed mapped surveys of their dominions. The best-known examples are perhaps Philip Apian's survey of Bavaria of the 1560s and Christopher Saxton's mapping of England and Wales in the 1570s. Alongside these better-known surveys, however, were numerous others of no less quality. They were undertaken, for instance, by Marco Antonio Pasi in the Este dominions of Ferrara and Modena in the 1570s, and by Egnazio Danti in the Papal States in the 1580s. Just prior to this Bologna, the home town of Pope Gregory XII (best known for his calendar reforms), had been surveyed. To commemorate this, and his own pride in his origins, the Pope in 1576 commissioned frescoes of the maps of the town and of the province of Bologna to decorate the walls of his dining and reception room in the Vatican.[44] As Francesca Fiorani has written, 'as large, beautiful frescoes these paintings delighted. As accurate images of the world, they instructed. And as both beautiful and scientific images they moved the cultivated papal court to the contemplation of the divine order of things'.[45]

The leaders of city states and provincial governments also projected their civic pride and more limited power through large maps that were intended for display. The city of Augsburg still owns a handsomely coloured copy of the Seld/Weiditz plan of the city of 1521, and the city of Venice still possesses – and treasures – the original woodblocks for De' Barbari's bird's-eye view of 1500. Many of these surveys remained in manuscript and were not intended for display. Printed or manuscript versions, however, adorned the walls of royal palaces, at a reduced scale and smaller size, with information that was considered confidential omitted and often with figurative propaganda elements added.

Sometimes a map could combine the historical, cultural and genealogical with the geographical and the scientific in the monarch's service. The larger manuscript model for John Speed's four-sheet map of England and Wales, engraved by Remond Elstrak in 1604 (p.62–3), is probably the 'kingdom of England drawn with the pen and coloured on a large table showing all the intestine wars besides where and at what place the battles were fought'[46] that was recorded as hanging in Whitehall Palace in 1613. The printed map combines information from Christopher Saxton's wall map of 1583 with depictions of failed attempts at invasion and the civil wars of past centuries, figures representing the different social classes of the land and a genealogical tree establishing the legitimacy of the newly installed Stuart dynasty.

Occasionally the maps on display might show lost lands which had belonged to ancestors but which the ruler still claimed, or even lands over which he aspired to rule, though they had never formed part of his or his predecessor's dominions. In a gallery in Hampton Court Henry VIII displayed maps of Normandy and of Scotland; he regarded himself as legitimate ruler of both, though Normandy had been lost three centuries earlier and no English king had been able to vindicate his claims to Scotland on a lasting basis. By contrast, in the 1570s the Duke of Savoy had a tapestry created from a specially commissioned map of the Netherlands.[47] As the husband of a Spanish princess, the tapestry would have served as a discreet reminder of the Duke's awareness that only a few lives separated him from dominion, as heir to the Spanish Habsburgs, over the Netherlands, and that he was already in a position to commission detailed maps of the region. More extravagant still are the claims embedded in the magnificent tapestry by Bernard van Orley, commissioned by the Portuguese king Joao III in 1525 to commemorate his marriage to a sister of Emperor Charles V. This showed the Portuguese king in the guise of Jupiter, with his sceptre placed above the position of Lisbon, guarding a globe dominated by the continents of Africa and Asia which had been allotted to his predecessors by the Papacy in the 1490s.[48]

The ruler of a minor state might compensate for his lack of military might by displaying maps that drew attention to his powerful allies, as well as to his personal sophistication and intellectual abilities. The decoration and a text panel in the Seld/Weiditz map of Augsburg referred viewers – and particularly its expansionist princely neighbours – to the protection that the

city enjoyed from the German Emperor. Duke René of Lorraine may well have subsidized the creation of Martin Waldseemüller's great world map of 1507, the first to use the name America, in order to demonstrate his intellectual pre-eminence.[49] Indeed, one result of the excitement generated by the Great Discoveries is that large sea charts and world maps became acceptable display objects in the grand surroundings of a palace, particularly if the presentation was elegant, the representation accurate and up-to-date and the information tilted to flatter the ruler whose palace it adorned (see pp.44–5 and 84–5).

28   See, for instance, the description by Juan Fernandez de Velasco of his reception at Whitehall in 1604, in William Brenchley Rye, *England as seen by Foreigners in the days of Elizabeth and James the First* (London, John Russell Smith, 1865), p.118.

29   See Simon Thurley, *Whitehall Palace. An architectural history of the royal apartments, 1240–1690* (New Haven and London, Yale University Press, 1999).

30   Molly Bourne, 'Francesco II Gonzaga and Maps as Palace Decorations in Renaissance Mantua', *Imago Mundi* 51 (1999), pp.51–81.

31   Cornelis Koeman and Marco van Egmond, 'Surveying and Official Mapping in the Low Countries, 1500–c.1670' in *History of Cartography* iii, pp.1252–3. The copy (1601) by Pieter Claessens of the original of 1571 is reproduced on p.1254.

32   Brenchley Rye, p.166; Hatfield House inventory 1611, cited in Geoffrey Beard (ed.), *Upholsterers and Interior Furnishing* (Newhaven and London, Yale University Press, 1997), pp.285–6. I am most grateful to Hilary Turner for kindly giving me this reference. Hatfield House was still a royal palace at the time when the inventory was compiled.

33   Celina Fox, 'George III and the Royal Navy', in Jonathan Marsden (ed.), *The Wisdom of King George III* (London, Royal Collection Publications, 2005), pp.303–7.

34   A splendid pair of terrestrial and celestial globes is still to be seen in situ in the Prunksaal, built in the 1720s, in the Austrian National Library in Vienna.

35   Accademia Toscana de Scienza e Lettere, 'La Colombaria', reproduced in Schulz, *Arte e Cartografia*, p.l.9.

36   *The Inventory of King Henry VIII. The Transcript*, David Starkey (ed.) (London, Society of Antiquaries, 1998), i, no.10752.

37   There were limits to what today would be called the 'spin': the walls of Whitehall Palace also contained maps which glorified the feats of his fellow princes and had been presented to Henry by their envoys. Not knowing where to store them, and perhaps not fully aware initially of their propagandistic purpose, Henry seems to have had them displayed where there was space on the walls of his palaces. Later he seems to have become more canny, re-using those that he could for his own propagandistic purposes.

38   See Barber, *Inventory* iii, forthcoming.

39   For detailed discussions of these cycles see Francesco Fiorani, *The Marvel of Maps. Art, Cartography and Politics in Renaissance Italy* (New Haven and London, Yale University Press, 2005); Francesca Fiorani, 'Cycles of Painted Maps in the Renaissance' in *History of Cartography* iii, pp.804–30 and its extensive bibliography. See also Juergen Schulz, *La cartografia tra scienza e arte. Carte e cartografi nel Rinascimento italiano* (Modena, Panini, 2006, revised edition of Italian translation originally published in 1990), which contains updated versions of important papers that originally appeared in English.

40   The complete decorative programme was never completed.

41   Jodocus Hondius's 'Christian Knight Map' of about 1597 could be interpreted as an answer in kind to the propaganda of the Vatican galleries – in this case showing the fight back of Protestantism with the background of an up-to-date and sophisticated world map. And see Peter Barber, 'The Christian Knight, the Most Christian King and the rulers of darkness', *The Map Collector* 52 (autumn 1990), pp.8–13.

42   Jean Boutier, *Plan de Paris dès origines (1493) à la fin du xviiie siècle* (Paris, BNF, 2002), p.95. A tapestry map of Leiden created in 1584 can still be seen in the Lakenhal in Leiden.

43   These, which include a map of the western Mediterranean, were first displayed in public in Westminster Hall in 1555 during the celebrations that followed the marriage of Mary Tudor to Philip II. For recent discussions see Hendrik J. Horn, *Jan Cornelisz Vermeyen: Painter of Charles V and his Conquest of Tunis* (Doornspijk: Davaco, 1989), *Der Kriegszug Kaiser Karls V gegen Tunis. Kartons und Tapisserien*, Wilfried Seipel (ed.) (Vienna, Kunsthistorisches Museum, 2000).

44   Fiorani, *The Marvel of Maps*, pp.143–57. The room also has an astronomical ceiling like that in Caprarola, the palace of the Pope's rival, Cardinal d'Este, but in this case it glorified Gregory.

45   Francesca Fiorani, *The Marvel of Maps*, p.154.

46   Brenchley Rye, p.165.

47   Inventory of the Vigilius de Ayatta collection, 1575, *Imago Mundi* v, pp.18–20. I am grateful to Guenter Schilder for this reference.

48   Jerry Brotton, *Trading Territories. Mapping the early modern world* (London, Reaktion, 1997), pp.17–18. The tapestry is reproduced on the cover and as illustration 2.

49   For this map, now owned by the Library of Congress, see most recently Toby Lester, *The Fourth Part of the World: the epic story of history's greatest map* (London, Profile, 2009).

THE GALLERY

In palaces maps were most frequently to be seen in the gallery.[50] This space evolved in Italy from the early fourteenth century and was initially no different from an open loggia, with which it continued to have close affinities. The essence of these spaces, which connected rooms in the ruler's private apartments, was their link between the interior and the exterior: the loggia was open to nature on one side; the gallery was lined with large windows on one – and sometimes on both – sides, from which the visitor would also see nature. Their luminosity made them ideally suited for the display of art works.

Renaissance theorists saw loggias and galleries (the terms were used interchangeably for a considerable period) as modern counterparts of the *ambulatoria* of Roman villas, which had been lined with paintings taken from nature. The accepted mural decorations for loggias and galleries, therefore, came to be depictions of nature – in the form of plants and animals but also, in perceived imitation of the Romans, of landscapes and townscapes. The earliest known examples – now virtually destroyed – are the views of Rome, Genoa, Florence, Venice and Naples in a loggia in the Belvedere villa in the Vatican. These were painted 'in the Flemish style' by Pinturrichio for Pope Innocent VIII between 1484 and 1487.[51] Bird's-eye views of towns had been fashionable for some time, and printed views of such towns as Rome, Constantinople, Pisa, Florence and Naples had already been published by the Florentine artist and mapmaker Francesco Rosselli.[52] By the 1540s, and probably considerably earlier, the printed and painted town views to be seen in galleries were being supplemented by other types of map, such as regional and world maps and decorative sea charts.

The galleries were also increasingly being used as spaces in which to display items from the rulers' collections. Paintings, particularly portraits, but also religious images, depictions of historic events and busts of great figures from history, joined and soon outnumbered the maps, views and natural history paintings. Rulers or their advisers also realized that such displays, ostensibly aesthetic and intellectual as they were, subtly reinforced positive messages about their own intellectual interests and abilities. This development had been foreshadowed a century earlier by the architectural theorist

The *Terza Loggia*, The Vatican

Bird's-eye view of Genoa in the palace courtyard, Viso del Marqués, Spain

Courtyard of the palace of the marqués of Santa Cruz, Viso del Marqués, Spain

Leon Battista Alberti who, in his influential *Trattato di Architettura*, encouraged the placing of elaborately framed paintings of classical triumphs, such as of Pompey's triumphs on land and sea, on the walls of galleries.[53] Such displays were particularly important for the rulers of small, economically weak and militarily insignificant states who could not make their mark on the world in any other way.

The nature of the gallery underwent a change as it moved to the less clement climate north of the Alps. Loggias were reduced in size, and galleries expanded as they developed into recreation spaces during periods of bad weather and (in the case of galleries situated next to studios) as places of mental relaxation. In the increasingly dynastic contexts of north European courts of the sixteenth and seventeenth centuries, the gallery's decoration, including the maps and views, tended to take on a strongly propagandistic tone, often including handsome versions of early national mapped surveys. These political notions fed back into Italy by the later sixteenth century. They led to the creation of splendid painted galleries, several of which, as we have seen, were strongly cartographic. Loggias also took on a more propagandistic role, as with the *Terrazza delle Matematiche* in the Uffizi in Florence and the *Terza Loggia* in the Belvedere villa in the Vatican. The increasing importance of maps as instruments of political propaganda could also be seen in Venice, where the number of maps displayed on the walls of the Doge's Palace also increased, despite a disastrous fire in 1577 which destroyed the earlier ones.[54]

Courtyards formed a kind of external gallery which were sometimes adorned with maps, particularly of towns. Specially commissioned views of towns in the dominions of the Austrian Habsburgs were created for the inner walls of the courtyard of the Palazzo Vecchio in 1565 as a compliment to the Habsburg archduchess Johanna on her arrival in Florence to marry the Duke's oldest son and heir, Francesco de' Medici. Between the early 1570s and about 1595 plans and views of European towns were painted above the grand doors in the ground and first-floor arcades lining the courtyard of the palace of the leading Spanish admiral of his time, Alvaro de Bazán, marqués of Santa Cruz, at Viso del Marqués in Spain.[55] They illustrated the owner's career and were copied from Braun and Hogenberg's multi-volume collection of town plans known as the *Civitates Orbis Terrarum*.

From 1600 Italian galleries increasingly became the first art galleries in the modern European sense of the word. In northern Europe, however, they remained primarily spaces for recreation and propaganda. There continued to be galleries in the private wings of palaces, and these sometimes contained expensive maps, as did the gallery of Queen Anne's husband, Prince George of Denmark (d. 1708) in Kensington Palace. Their counterparts in what were now the semi-public parts of the palace grew ever grander. Several European rulers, such as the king of Sweden, followed the example of Louis XIV at Versailles, where the *Galerie des Glaces* increases the luminosity of the magnificent interior with a series of enormous mirrors facing the large windows. These galleries often lead to the ostensibly private but, by that date, semi-public spaces, such as the bedchambers and dining rooms where rulers received their more prestigious guests. By this time, however, the grand galleries do not appear to have contained any maps.

50   For the following section see Claudia Cieri Via, '"Galaria sive loggia": modelli storici e funzionali fra collezionismo e ricerca', introduction to Italian translation of W. Prinz *Galleria* (Modena, Panini, 1988), pp.vii–xxiii; Wolfram Prinz *Die Entstehung der Galerie in Frankreich und Italien* (Berlin, Gebr. Mann, 1977).

51   Schulz, '*La veduta di Venezia di Jacopo de'Barbari: cartografia, vedute di città e geografia moralizzata nel Medioevo e nel Rinascimento*', in *La cartografia tra scienza e arte*, p.48.

52   Roberto Almagià, 'On the Cartographic Work of Francesco Rosselli', *Imago Mundi* 8 (1951), pp.27–34.

53   Quoted in Cieri Via, p.xii.

54   Schulz, *Cartografia tra scienz e arte*, pp.91–139.

55   Juan del Campo Munoz, *Breve Historia del Palacio de Viso del Marques* (Madrid, Museo Naval, 1994), pp.20–23, 32–5.

Henry VIII's map of Italy

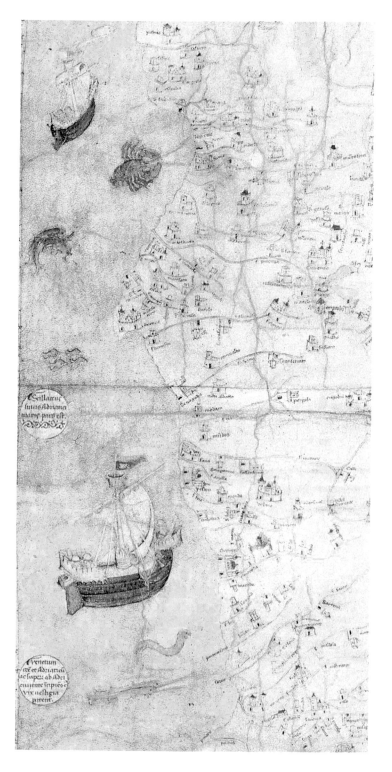

This hitherto little-studied manuscript map of Italy is probably the earliest surviving 'modern' map of the region. Its handsome, though now rather battered, form suggests that it was intended for an important recipient. It could well be the 'Discription of Italie of parchement sette in a frame of woode' in Whitehall Palace that is mentioned in inventories of Henry VIII's goods dating from the 1540s. It has nail holes along its edges, showing that it was once framed, and it forms a part of the collection of Sir Robert Cotton who acquired many maps and manuscripts that had once been owned by Henry VIII. Cotton's collection of manuscripts became one of the British Museum's foundation collections in 1753 and was transferred to the British Library on its creation in 1973.

The map bears traces of its original splendour, as can be seen in the gilt lettering in a courtly gothic hand on a background of blue lapis lazuli at the top of the map, and the decoration of fishing boats and sea animals surrounding the coastlines. It is oriented to the south, as many early maps were, and on the basis of content and handwriting can be dated to the second quarter of the fifteenth century.

Close examination reveals that the map is a combination of several elements. On the one hand it consciously incorporates traditional values, reinterpreted in a humanistic manner characteristic of the early Italian renaissance. The broad lines of the coastal outline are derived from a map of 1335 by the Venetian Fra Paolino Minoritas. The text on either side of the image incorporates geographical information about Italy from Ptolemy's *Geographia* and from the writings of Pomponeus Mela and Isidore of Seville, as well as a summary of the anonymous *De origine urbium Italie*, a history of the towns and cities of the peninsula. Thought to have been written by a north Italian lawyer in the late fourteenth century, the narrative highlights the antiquity and cultural pre-eminence of Italy, and the shared history of the Italians going back to the start of time. The wording of the original text, however, was amended on the map so that it finished with the foundation of Venice and emphasized that the future of Italy lay with Venice, its only unconquered city. This is not surprising since the map shows unmistakable signs, not least in its use of dialect, of having been composed in Venice.

The map is, as its title suggests, modern. The treatment of the coasts shows the influence of recent sea charts, while the depiction of the interior of Lombardy, the Veneto and central Italy is copied from detailed surveys, at differing scales, commissioned for military and administrative purposes since about 1290, particularly by Venice and the duchy of Milan. In places one can see the break between several different surveys. Southern Italy has very few names because it was culturally and economically far more backward. There was no competition between rulers for resources and the military surveys that characterized northern and central Italy had not yet been commissioned. This map, one of the first to bring all the information about Italy together on a peninsula-wide basis, also served as a visualization of a common Italian cultural entity.

The map would have been regarded as outdated inside Italy once a new and updated image of the peninsula, in part derived from maps like this, had been created for public rooms in the Doge's Palace in Venice by Antonio Leonardi in about 1459 (see pp.48–9). However, this old image was presumably regarded as still being acceptable for use as a diplomatic gift. It was surely welcomed when presented, probably by a Venetian ambassador, to Henry VII (reigned 1485–1509) or to Henry VIII at the start of his reign in 1509. After 1494, when a French invasion unleashed the Italian Wars, Italy became the fulcrum of European politics. This map would have enabled the English king, and his advisers and courtiers, to follow the course of the Italian campaigns far better than any printed map that was available before 1515.

Milanesi (2007–8), pp.153–76; Barber (2009), pp.105–9.

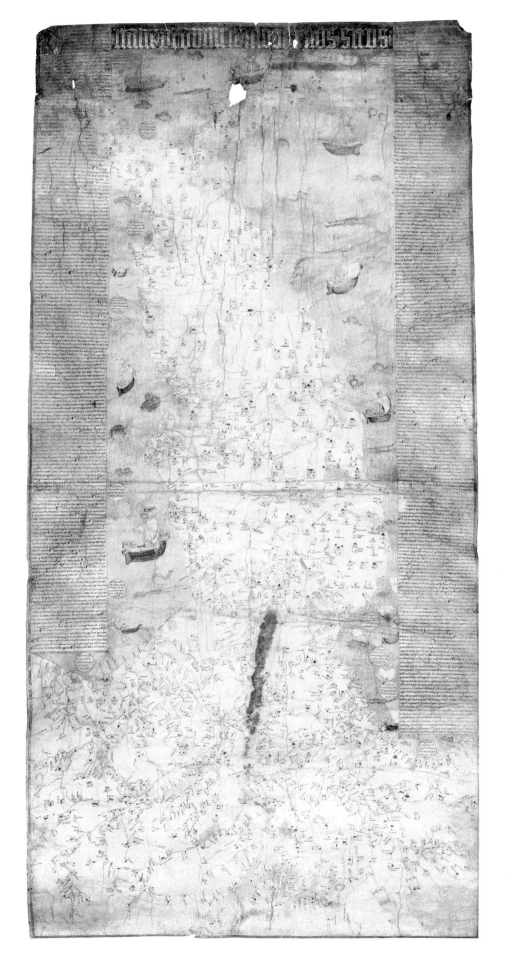

Anon., *Italie Provincie Modernus Situs*, Venice, *c.*1425–50

Manuscript on vellum, 137 x 65 cm

British Library Cotton Roll xiii

An image to stick in the mind

The magnificent printed plan of Venice published in 1500 is a celebration of the power of the Venetian Republic, and a symbol of the continued maritime, commercial and artistic successes upon which production of such monumental works depended. Its dramatically elevated viewpoint, large scale and symbolic meaning draws heavily on the ideas behind Italian painted galleries of towns, but, crucially, its production as a wood-cut enabled its statement of Venetian glory to be diffused throughout the gallery spaces of Europe. For example, the print is recorded to have been in the collections of Henry VIII and Ferdinand Columbus and its survival in some of the major collections today is testimony to its far-reaching appeal.

Unlike Rome, the city of Venice did not have a heritage that a European ruler wishing to associate his or herself with the Roman Empire and antiquity could adopt. Nevertheless, by manipulating its image through paintings and maps, by controlling this through printing privileges and by demonstrating the special, unique favour the city enjoyed through adapting the benevolent figures of Mercury and Neptune for its purpose, the Venetian state could surround itself with a timeless and attractive mythology which demanded admiration from its allies and enemies. The map is not only an enduring demonstration of Venice, and a striking adver-tisement for the city, but also undoubtedly a work of art. It must have appealed to the cultured, art collecting rulers of Europe, who placed the map in their galleries as a counterpoint to representations of their own towns and military glories. The enduring greatness of Venice thus came to be symbolized in the palaces of European power.

Prints as large as this had rarely been attempted before 1500, and no earlier examples have survived. The effort and technical skill required to produce the print meant that it would have been prized in a similar way to a painting or fresco. The survival of the woodblocks used to print the map suggests an interest in the production process, in addition to the finished product. Jacopo de' Barbari (c.1440–1515), influenced by his meetings in 1495 with the German artist Albrecht Dürer (1471–1528), compiled the map over three years by piecing together individual drawings and sketches. Unsurprisingly it was an expensive map, and at 3 ducats affordable only to the wealthiest, not to mention the most tasteful, of art collectors and rulers. De' Barbari's fame preceded him north of the Alps, and he was involved in producing huge celebratory woodcuts for Emperor Maximilian I in 1504.

Schulz (1978), pp.425–74.

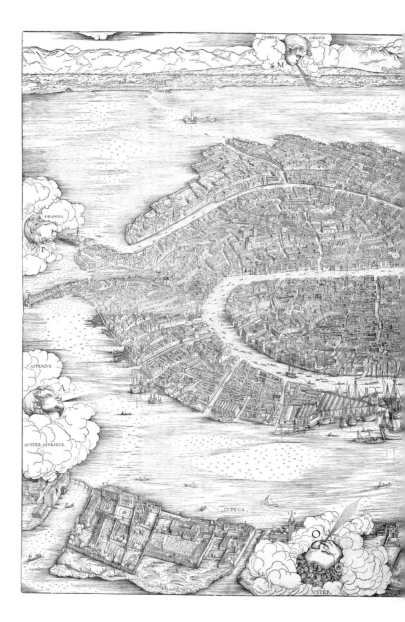

Jacopo de' Barbari, *Venetie MD*, Venice, 1500
Woodcut on six sheets, 133 x 281 cm
British Museum, Department of Prints and Drawings, 1895, 0112.1192–1197

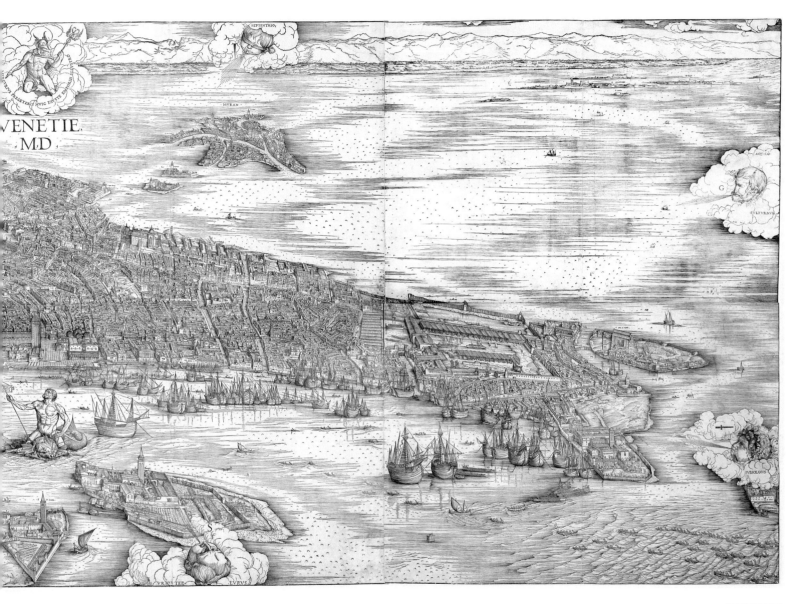

Defence into attack

This map of 1530 presents a visual record of the advance of the Ottoman army in Eastern Europe during the 1520s, which had culminated in the unsuccessful siege of Vienna in 1529. It is a map of a theatre of war, but the symbolism embedded within it reads as a celebration of victory, not only for the Habsburg monarchy but also for all of Christian Europe. The significance of such triumphal maps in the gallery of a victor would not have been lost on the viewer. They served as reminders of the military might and prowess of the ruler and his allies who displayed them.

The siege of Vienna by the Ottoman army under Suleiman I (1494–1566) constituted the high watermark of westward Turkish expansion in Europe. Such expansion had already lasted a century and had most recently included the capture of Belgrade in 1521 and the partitioning of Hungary in 1526. Inclusion of this string of events is the reason for the map's extensive geographical coverage. By incorporating most of Europe, the six-sheet map is able to illustrate not just the siege but also the events leading up to it – including the taking of Buda, the long march of the Ottoman army from Constantinople and Bulgaria, and its bombardment of what was then the major Hungarian city of Bratislava (Pressburg). The map survives in this one example, recently discovered in the collection of the Earl of Leicester.

The outcome of the siege of Vienna was clearly of Europe-wide importance – Henry VIII is documented as having a representation of the siege in his gallery. This may explain the decision by the mapmaker Johann Haselberg to produce a printed map for a wider audience, rather than a manuscript map for a single patron. The arms of fourteen European monarchs are displayed along the top of the map, while the monarchs themselves watch the progress of the siege from their respective countries. Haselberg's map, however, is more than simply a record of a military triumph: it turns defence into attack. The map was printed in at least two editions by the Nuremberg publisher Christoph Zell. On both occasions it was accompanied by a booklet that put forward ideas for a Christian reprisal in the form of a crusade against the Ottoman threat. The Muslim crescent above Jerusalem, recaptured by the Ottomans in 1517, would have been of further significance as a motivational tool for the advocates of such an expedition.

Meurer and Schilder (2009), pp.27–42.

Johann Haselberg / Christoph Zell, *Descriptio Expeditionis Turcicae Contra Christianos Anno Domini XXIX*, Nuremberg, 1530

Woodcut on six sheets, 62 x 99 cm

Collection of Viscount Coke, Holkham Hall

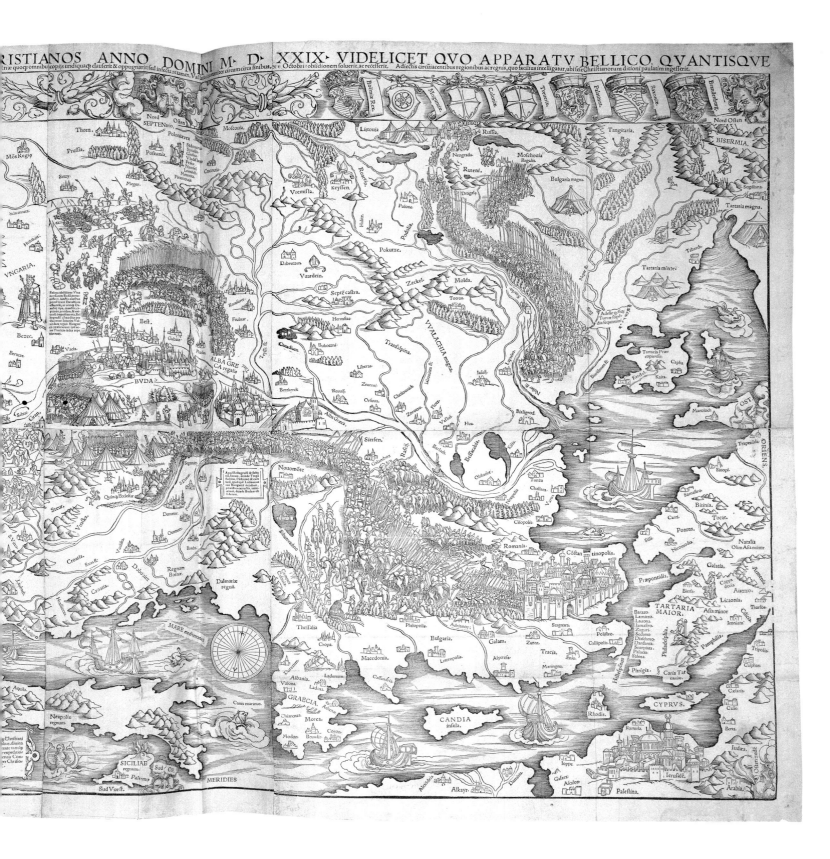

RISTIANOS ANNO DOMINI M·D·XXIX· VIDELICET QVO APPARATV BELLICO, QVANTISQVE

All that glitters …

Diogo Homem (*c*.1520–aft.1575) was the son of Lopo Homem, the leading Portuguese mapmaker of the mid-sixteenth century and the head of the main Portuguese mapping establishment, the *Armazém de Guinéa*, in the years when Portugal led the world in mapping shores beyond Europe. In 1544, as a result of his involvement in a murder, Diogo had to flee Portugal. He was warmly welcomed in England because of his outstanding mapmaking skills and the strategic importance of the geographical information that he carried in his head. After some years in England, where he created a handsome atlas for Mary I (now British Library Add. MS 4515A), Homem settled in Venice.

This chart was created towards the end of his life. It is very different from charts that he had made throughout most of his career, both in appearance and subject-matter. Stylistically the chart lacks the flamboyance of his earlier work, which abounded in colourful banderoles and vignette town views. Furthermore, Homem's previous charts had ranged throughout the world. This one is confined to an area that had been regularly mapped for the previous 250 years and some of his outlines, for instance those of the British Isles, were seriously outdated. For all that, there is an unmistakable glamour about the chart. The simplicity of its outlines contrasts with the gold leaf used for the coastlines, the ornate windroses and the well-executed cherubs' heads, which serve as windheads and probably were drawn by professional artists.

The provenance of this chart is not recorded before the middle of the nineteenth century, and it is not known why or for whom it was created. It could, however, only have been for display, possibly in a Venetian patrician's palace lining the Rialto. It may have been linked to an attempt by Homem to win employment from the Venetian senate as successor to the recently deceased chartmaker Battista Agnese, whose style he imitates and who seems previously to have worked, possibly informally, as chartmaker to the Venetian republic.

Cortesao and Texeira (1960), ii, p.39, pl.146; Barber, *The Queen Mary Atlas* (2005), pp.34–8.

Diogo Homem, [*Chart of the Mediterranean and western coasts of Europe*], Venice 1570
Manuscript on vellum, 74 x 114 cm
British Library Eg. MS 2858

An intimidating map

European rulers frequently tried to intimidate visiting envoys by displaying detailed maps of their masters' lands and forts, with the implication that the maps were a first step towards conquest. Bird's-eye views of French towns being successfully besieged by the English, for instance, decorated the walls of a specially built pavilion at Greenwich during festivities for the French ambassador in 1527.

Few maps could have had a blunter message than this. The dukes of Savoy, whose states straddled both sides of the Alps, had always had a difficult time in steering a course between powerful neighbours. Its dukes had tried to play off the great powers, and particularly France, which lay to the west, against the Holy Roman Emperor (Germany) and Spain, which owned extensive lands in Italy to the east. No Savoyard ruler had played this game better than Victor Amadeus II (reigned 1675–1730), who became notorious for his opportunistic switching of alliances. In 1703–4, during the War of the Spanish Succession, however, he deserted Louis XIV of France and allied with the Holy Roman Emperor, Leopold I, before German troops were close enough to Turin to support him against the French. Louis XIV decided to occupy Savoy on a permanent basis and to force its perfidious ruler into exile. French troops invaded from west and east, and by the middle of 1704 the Duke had lost almost everything.

This map represents the French cartographic view of the situation. At first glance the map, designed by one of the royal geographers, seems uncontroversial. Based on the detailed 15-sheet map of the Savoyard lands by Giovanni Tommaso Borgonio that had been published in 1680, it shows the states ruled by the Duke on either side of the Alps. It highlights the political structure of the region, which included numerous papal and imperial fiefs in the hills bordering the plain of Lombardy. This depiction is presumably taken from the information contained in the 'memoirs from Turin' mentioned in the title. In keeping with the belligerent nature of those years, the key at the bottom is flanked by the war gods Mars and Pallas Athena and surmounted by the arms of the king and dauphin of France, while a cherub carries the laurel of victory above Mars. It is the imagery at the top-right, however, that would have most alarmed any Savoyard official who saw it. Above the title panel, where one might have expected to see a portrait of Victor Amadeus, looms Louis XIV or 'Louis the Great' (*Ludovicus Magnus*).

Simcox, (1983), pp.134–50.

Jean Besson, *L'État du Duc de Savoie de ca et de la des Monts ... dressé sur des mémoires envoyez de Turin*, Paris, 1704 [1709]

Copperplate engraving on six sheets, 130 x 110 cm

British Library Maps K Top. 76.9 – 2 TAB

Empire's attributes

It is well known that Renaissance rulers liked to display maps and views of their most important towns and cities. Philip II of Spain (1527–98), for example, is known to have commissioned Anthonis van den Wijngaerde (d.1572) to create a series of images of major Spanish cities in the 1570s for precisely this purpose. The most suitable format for displaying a town view to fit into the space in a gallery was the profile or long panoramic view. Its narrow format enabled the artist to portray the vertical as opposed to linear mass of a city, along with which an expansive foreground scene of human endeavour could be incorporated.

This panorama of Seville by the Dutch artist Simon Frisius (c.1580–1623), published in 1617, remains a monument to the importance of the town for the Spanish Empire. As the city from which Spain's American empire was administered and the main port of the New World fleet, Seville was not only home to a flourishing chart-making community, it was also the point of arrival for riches obtained in the Spanish colonies of South America. Having been offloaded, this gold was counted and registered in the *Torre del Oro* (Golden Tower), which stands prominently in the middle right of the engraving. Many signatures of Spanish colonial power are apparent: the cannons in the harbour waiting to be loaded into galleons; the Christian iconography on sails, flags and in the streets; and the dominant churches accentuated against the skyline. These visualize the religious zeal which accompanied the foreign policy of Charles V (reigned 1516–55) and his successors.

The production of multi-sheet engraved panoramas flourished in Amsterdam during the seventeenth century, but it is unusual to find a complete set of sheets today. The Seville view is one of only three surviving examples. The panorama has been trimmed at the top and bottom edges, and so has lost the title and letterpress description of the town below. It is made up of sheets which were printed at different times and fitted together in the early eighteenth century.

Collijn (1915), no.82, p.27.

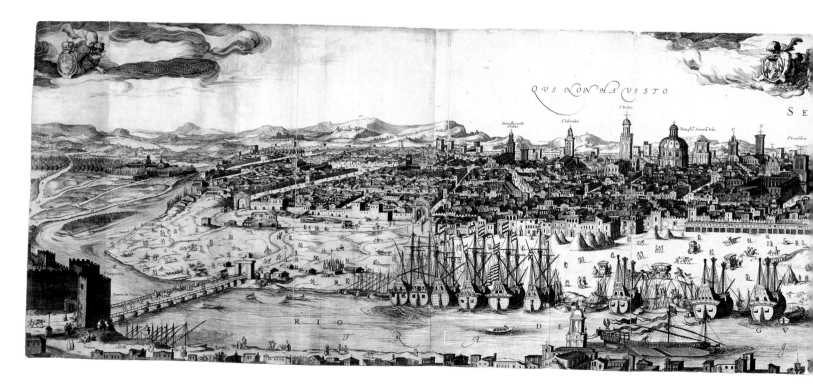

Simon Frisius / Johannes Janssonius, [*Hispalis vulgo Seviliae urbis toto orbe Celebrissimae Hispaniaeque Primariae Effigies*], Amsterdam, 1617

Copperplate engraving on four sheets, 49 x 225 cm

British Library Maps K.Top 72.16

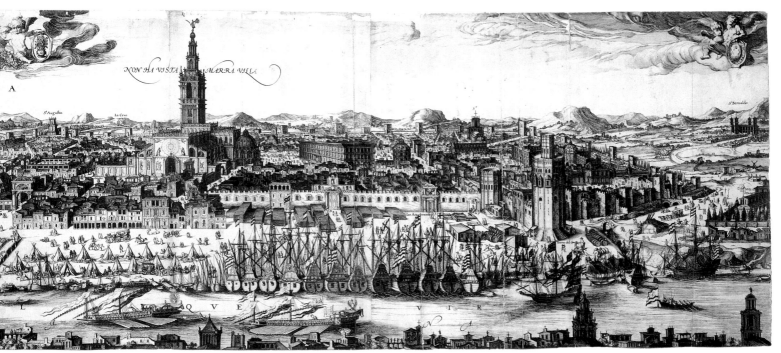

A Medici beauty

In 1576 Grand Duke Francesco de' Medici of Tuscany summoned to Florence a Benedictine monk, Stefano Bonsignori, from the Monte Oliveto monastery near San Gimignano. Francesco had fallen out with Egnazio Danti, another monk and a talented cartographer who had initiated an ambitious mapping project in the *Guardaroba Nuova* in the ducal palace, now the Palazzo Vecchio (p.82). As a result, Francesco wanted Bonsignori to complete the job. By the early 1580s Bonsignori was surveying the whole of the Medici dominions, leading to the publication of detailed maps of the provinces of Florence and Siena, and of this map of the city of Florence in 1584. In 1589 these provincial surveys were the sources for the painted maps, created by the artist Leonardo Buti, which still adorn the *Sala delle Carte Geografiche,* also known as the *Terrazza delle Matematiche* in the Uffizi.

A magnificent view of the city could be enjoyed from the loggia, so the map of Florence did not need to be turned into a painting. The printed version of Bonsignori's survey on nine sheets, however, became much sought after. It rendered obsolete the previous standard cartographic image engraved by Francesco Roselli in the 1470s. Bonsignori's map probably found its way into princely galleries throughout western Europe as had its fifteenth-century predecessor, an example of which was displayed in Whitehall Palace in the 1540s. As the title implies, it was the first 'accurate' – that is, mathematically surveyed – town plan of the city. Bonsignori portrayed himself in his robes at the bottom left of the map. In contrast to the earlier map of Florence, on which the mapmaker was depicted as an artist, Bonsignori shows himself with his surveying instruments – as though making the point that local mapmaking or chorography was now the business of the professional surveyor and no longer of the painter.

The map of Florence is aesthetically pleasing, well engraved (by another monk) and still manages to show the facades of individual buildings, albeit from a vertiginous angle, in such a way that the street layout can be seen. The original dedication to the Duke (missing from this particular edition) stated unequivocally that the map's function was to show 'the ornaments [to Florence] created by Your Highness, your father and your forefathers', meaning that it was intended as much to celebrate the cultural and scientific achievements of the Medici as to provide practical information about the city. The printed map thus shared the same objective as the painted cartographic projects also undertaken by Bonsignori and by Danti in Florence.

Indeed, in some respects the political objective got the better of the scientific. Bonsignori optimistically showed some features that did not yet exist, such as the fountain in Piazza Santa Croce, as well as others, like the column to peace in the Piazza San Marco, that were never built. He also showed, though, new features, mainly constructed under Cosimo I. These were hydrographic as well as architectural. Water was such an important commodity in a Mediterranean city like Florence that an influential tract published in 1510 recommended that hydrographic plans were suitable for display in cardinals' galleries. So, in addition to showing the Uffizi Palace, the Boboli Gardens and Ammanatti's *Fontana di Nettuno* in the Piazza della Signoria, Bonsignori's map also shows newly constructed channels in the Arno, with rows of trees strengthening the embankments and the numerous public fountains created under Cosimo. A wider context is hinted at by the miniature eastern hemisphere to be seen in the middle of the compass rose.

The map was reprinted in 1594 with minor additions (such as the fortress of the Belvedere), and again as late as 1660.

Else (2009), pp.168–85.

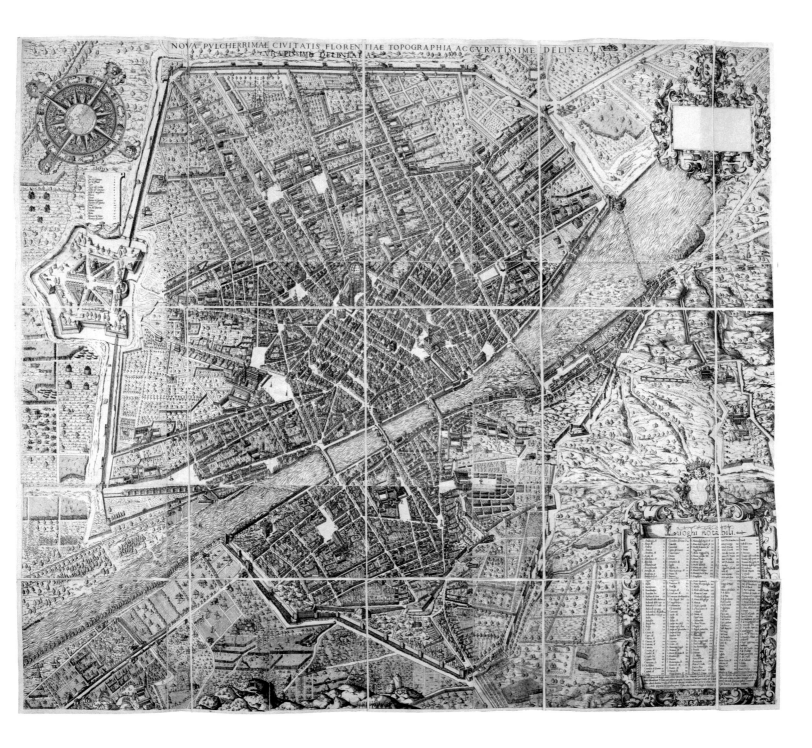

Stefano Bonsignori, *Nova pulcherrimae civitatis*
Florentiae topographia accuratissime delineata
1584 (Rome, Rossi, 1660)
Copperplate engraving on nine sheets, 125 x 138 cm
British Library Maps * 23480. (2.)

World power by association

Among the maps hanging in the Privy Gallery in Whitehall during the sixteenth century was a 'large Mappe of the whole worlde of parchement sette in a frame of woode havinge the kynges armes therin'. Probably presented to Henry VIII at some time between 1524 and 1526 by Girolamo Verrazzano, brother of the more famous explorer Giovanni, it was still on display at the end of the century. It may have borne a passing resemblance to this manuscript world map.

The title claims that the map shows 'everything that has been discovered on a flat sea chart'. It strives to appear accurate and up-to-date but honest. Though the mapmaker correctly suggests that there was a strait separating America from Asia (top right), for instance, he admits that what is now Alaska is 'terra inchognita' [*sic*.]. Though several ships and fabulous sea monsters are to be seen, they are not particularly prominent.

This is not a nautical chart in the true sense. The mapmaker, Antonio Millo, whose name appears in the banderole at the bottom centre, does not superimpose the outlines of the world on a network of rhumb lines (a linear depiction of wind directions) as he would have done had he intended to create a functional chart for use at sea. Instead, in scientific fashion, he provides the map with a grid of parallels and meridians and marks the degrees of longitude and latitude in the margins. He places the zero meridian, near the Azores (as was then customary), indicating its position by adding a scale of longitude to it. The interiors of the continents receive far more attention than was usual on sea charts, with numerous regions, towns, rivers and deserts being named.

Millo has tried to work only from reputable sources, but he does not seem to have employed the most-up-to-date maps published in Antwerp, although they would have been available in Venice. Instead, probably through laziness or misguided patriotism, he relies on Italian maps that had been published twenty or more years earlier and particularly those of Jacopo Gastaldi (1500–66), who had been cartographer to the Venetian Republic from 1548 until his death. The depiction of Scandinavia is copied from a map by the Flemish cartographer Cornelis Antoniszoon, which had been reprinted by Gioan Francesco Camocio in 1562. Where Millo did not have more up-to-date information he relied on the outlines to be found on traditional sea charts, some of which, in the case of Great Britain, for example, went back to the early fifteenth century. Most of the Pacific is excluded because, despite the circumnavigations of Magellan and more recently Drake, it was largely unexplored.

Political information is extremely sparse – with one big exception. By means of small circular windroses and strategically placed Spanish and Portuguese coats of arms, Millo highlights the Line of Tordesillas. This had been drawn by the Pope in the 1490s to divide the world outside Europe between the Portuguese and the Spanish. The line was still significant in the late sixteenth century, even though there had been a personal union between the two kingdoms since 1580 when Philip II of Spain acceded to the Portuguese throne, following the extinction in the male line of the Portuguese kings of the Avis dynasty.

Antonio Millo (fl.1567–91) was a Greek sailor from the island of Milos who had probably trained in Venice with the exiled Portuguese cartographer Diogo Homem (p.34). In the year he created this map Millo was appointed *Amiraglio*, or chief pilot, in the Venetian ports in the eastern Mediterranean. It is possible that the production of this map smoothed his path to that appointment. The as yet unidentified recipient, probably a nobleman from the Venetian mainland with a home near the Habsburg dominions to the north, must have been flattered. The depiction of his arms, occupying the position of honour at the bottom left, associate him, as the English royal arms on Verrazzano's map had Henry VIII, with the most powerful monarchs in the world, whose arms are shown elsewhere on the map, and with the world itself – or, at least, with knowledge of it.

Tolias (1999), pp.39–41, 191–200.

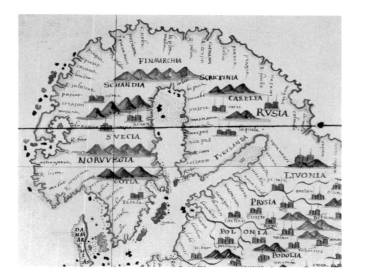

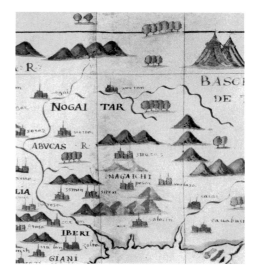

Antonio Millo, *Tvto el discoperto in carta marina in piano*, Venice, 1582
Manuscript on vellum, 119.5 x 220 cm
British Library Add. MS 27470

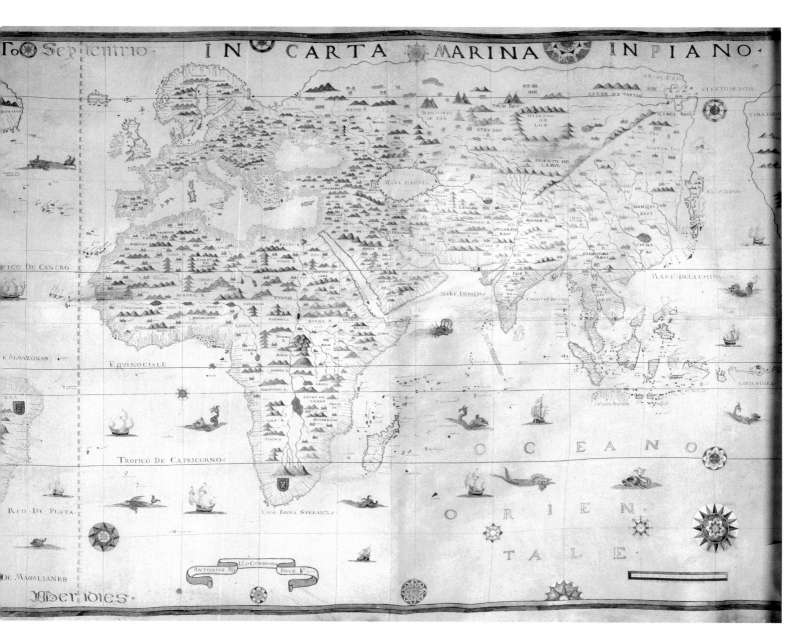

Enlightened colonialism

Just as the gallery offered opportunities for a ruler to display maps of his own lands, chief cities and glorious victories, it also offered the chance to show recent additions to his empire, and even land that had not yet been conquered. The perceived realities of these far-flung places were portrayed in a positive way for a ruler and his visitors, who were often never to see them with their own eyes.

This map of Brazil is effectively a compilation of maps, sketches and paintings made during the Dutch administration of Brazil from 1630, at the behest of the Governor-General of Dutch Brazil, Count Johan Maurits of Nassau (1604–79). Maurits has been likened to a Renaissance prince in his thirst for knowledge and his promotion of the arts and sciences. His employment of a number of scientists and artists to record and research the geography, terrain, wildlife, people and customs of the newly claimed Dutch territory was no doubt motivated primarily by security and trade, but perhaps also by curiosity and the hope of personal immortality.

The compiler was Georg Marcgraf (1610–43), a geographer and natural historian who had worked for Maurits in Brazil. It is uncertain whether Marcgraf himself was responsible for the mapping of the Brazilian coast, or whether he updated existing charts. What is more certain is his association with artists such as Albert Eckhout (c.1610–65), who produced ethnographic portraits and drawings of natural objects collected by Marcgraf. The source for the the depiction on the map of animals, which include a sloth, an anteater and a boa constrictor, as well as the several expansive scenes of Brazilian life, are thought to be sketches made by another artist, Frans Post (1612–80). Post's scenes include a sugar mill, plantation buildings, an ostrich hunt, fishing and feasting and a battle. Two indigenous tribes are afforded special prominence. The idealized depiction of the Tupinambu settlements with the supervised tribesmen working for their Dutch overlords – slavery, to give it its proper name – would have contrasted with the vignettes emphasizing the barbarism of the customs of the Tapuya people.

Even supposedly neutral documentation contains an agenda. Marcgraf's portrayal of the reality of the Dutch empire in Brazil, rendered with consummate skill on the map and with a large explanatory key, would have offered proof of the origin of the exotic fruits, vegetables and other crops and of the money that were flooding into the ports of Amsterdam and Rotterdam. The map remained the most concise representation of Brazil for much of the seventeenth century. Most of Post's Brazil work on which Marcgraf relied so heavily, is lost, though a volume of his drawings, now in the British Museum, formed the basis of illustrations in the 1647 *Rerum per Octennium in Brasilia ... Historia*, a history of Dutch Brazil under Maurits. The author of the *Historia* was Caspar van Baerle (1584–1648) and the publisher Joan Blaeu (c.1599–1673). A four-sheet version of Marcgraf's map had already been published in 1641 by van Baerle, and it was from these four sheets that Blaeu created the present wall map that same year.

It is a reflection of the propagandistic intent of the Marcgraf map, as well as its obvious vulnerability as an artifact, that three of the remaining copies survive in large, bound atlases prepared for illustrious royal patrons. The Dutch merchant Johann Klencke included this example in the outsize atlas that he presented to Charles II upon his restoration to the English throne in 1660 (see pp.92–3). Another example was presented by Maurits himself to Frederick William, the 'Great' Elector of Brandenburg.

Whitehead and Boeseman (1989); Whitehead (1982), pp.17–20.

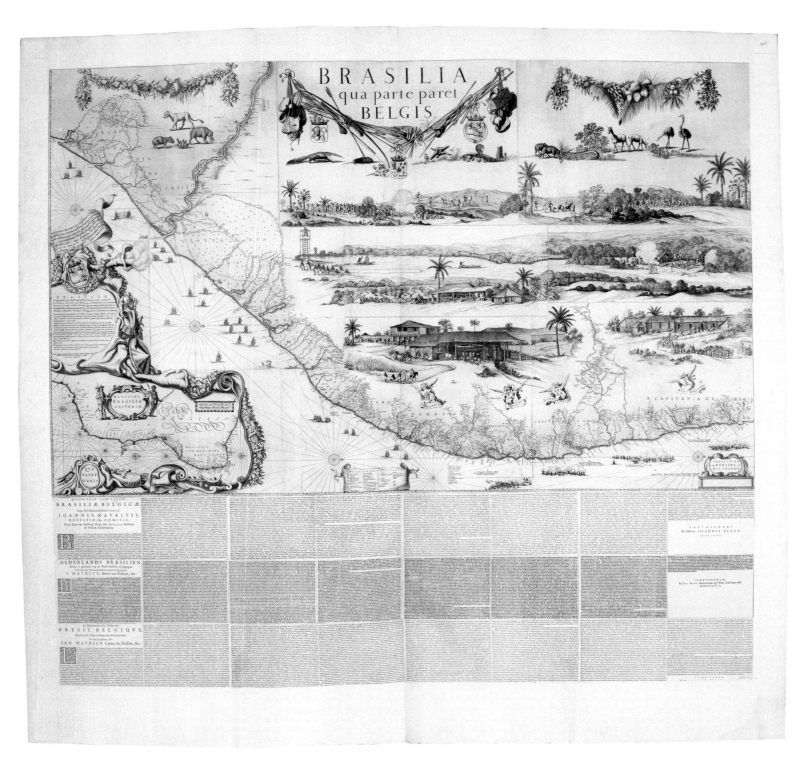

Georg Marcgraf / Joan Blaeu, *Brasilia qua parte paret Belgis*, Amsterdam, 1647
Copperplate engraving on nine sheets, 120 x 155 cm
British Library Maps K.A.R.(38.)

A conquered kingdom

Galleries were often used by rulers to display objects from their own collections. George of Denmark (1653–1708), consort to Queen Anne of England, is known to have used his private gallery at Kensington Palace for this purpose, and this presentation copy of Henry Pratt's wall map of Ireland, dedicated to George himself, would almost certainly have been displayed at Kensington. The map had been advertised in the *London Gazette* and *Daily Courant* in April 1708, priced at an expensive 1 guinea, but this example is obviously an embellished version, prepared specifically for George in colour and with silk trim. It would have been given in the hope of patronage, good fortune and further commissions.

George of Denmark was a highly suitable patron for Pratt's map. As Lord High Admiral and General of Her Majesty's forces, he would in theory have appreciated the attention Pratt had paid to roads, barracks and the locations of a large standing army in what was seen as a conquered kingdom, yet one still susceptible to foreign influence. French troops had attempted to land in Scotland in 1707, and the memory of the Glorious Revolution still lingered. The Irish war that followed the Revolution is subtly referred to in the plan of Drogheda, one of the town plans flanking the map, which includes a plan of the Battle of the Boyne. This would particularly have pleased George, a staunch Protestant who had supported both William's campaign against James II and his accession to the English throne in 1689. Whether George fully understood the map's subtle message of peace underlined by tight military control is uncertain, however, as Charles II is alleged to have said of him to Anne, 'I have tried him drunk and I have tried him sober, and there is nought in him'.

Henry Pratt was an estate surveyor who worked in both Ireland and London. The town plans flanking his map are based on unpublished military plans owned by the 2nd Duke of Ormonde, which suggest that Pratt held an official position within the Irish administration. The central portion of the map is a combination of the survey of Sir William Petty of 1675 and improvements made by Pratt himself.

Andrews (1997), pp.153–84.

Henry Pratt, *Tabula Hiberniae Novissima et Emendatissima…*, London, 1708
Copperplate engraving on four sheets, 140 x 138 cm
British Library Maps K.Top 51.18.TAB END

The sport of kings

Hunting, 'the sport of kings', demonstrated a ruler's virility and leadership qualities, so bird's-eye views of palaces set amid royal hunting grounds were a popular theme for display in the galleries of royal palaces. Henri II and Henri IV of France had even commissioned galleries entirely filled with painted plans of French hunting forests for the royal palace at Fontainebleau.

This manuscript map shows the spot which George I, a rather shy and reserved man, loved above all others as the only place where he could briefly escape from the cares of office. The Goehrde was the main hunting forest in his German dominions and had a European-wide reputation. George visited whenever he could, usually for eight weeks a year between mid-September and mid-November, a practice that he continued even after becoming king of Great Britain in 1714. He commissioned Louis Rémy de la Fosse to design an extensive new hunting lodge there in French style. Effectively a palace, the lodge was built between 1706–9 with a sizeable visitors' wing, stables for over 1,000 horses, coach houses, a wine cellar, a bakery and a theatre. These were all needed for the great hunts, which were important social and diplomatic events. As well as spending time with his brother and legitimate and illegitimate daughters, George would meet other north German rulers at the Goehrde. There would be much carousing and singing of hunting songs, but also theatrical performances. In the late afternoons and evenings, George would reluctantly transact business with his ministers and occasionally treaties would be signed.

This anonymous map of the forest, drawn at a scale of 1:10379 (about 10 centimetres to a kilometre, or just under 6 inches to the mile), gives all the information required to organize a good hunt. As well as showing routes, rides, ditches, fences and places to change horses and to store carcasses inside the main forest, it also indicates routes through the neighbouring countryside and identifies the different parts of the palace complex. The decoration and the minutely detailed draftsmanship, however, transform the map into a celebration of King George as he would have wished to be seen: as a huntsman and builder. He is portrayed at the top in profile, after a model by Kneller. At the top right is a view of the facade of the palace, where even the carvings in the pediment can be made out. To the left is a rear view of the main palace, with the detached theatre to the left. These seem to be the best surviving records of the appearance of the palace, which was demolished in 1829. On the lower right side the hunt is shown setting out, with the palace in the background, and at the left is the hunt itself. The king can be recognised even from behind by the Garter ribbon across his shoulder. A panel at the centre contains a glorification of the hunting dogs and hunting. The reality was rather different from the image, however: in the course of a hunting season no more than 36 deer were ever killed, and sometimes the number was as low as six.

The Goehrde thrived under George I and George II, but under George III, who never visited Hanover, it was leased out and declined.

Prüser (1969), pp.58–72; Hatton (1978).

Anon., 'Plan de la Forest de Goerde, et de la maison de chasse de S.M. le Roy
de la Grande Bretagne, dans ses etats de Bronsvic-Lunebourg; representant
en particular les parties et cables du Goerde et les routes pour la chasse
qui les traversent', Hanover, 1717

Manuscript, 110 x 141 cm

British Library Maps K. Top. 100.9–1

THE AUDIENCE CHAMBER

The great halls of medieval kings, the halls or audience chambers of city states such as Venice or Augsburg, and even the antechambers through which visitors had to pass beforehand, were also important locations for large wall maps. Equivalent spaces existed in the palaces occupied by the governors or the estates of provinces, and in the town halls of large cities which did not enjoy self-governing status.

Henry III is recorded as commissioning a large 'mappa mundi' – probably but not certainly a world map – for the great hall at Winchester in 1239.[56] World maps seem to have long continued to be the type of map most commonly seen in or near audience chambers and their equivalents, though the extremely sparse surviving written and visual evidence is almost entirely confined to Italy. Doge Francesco Dandolo was said by the chronicler Morosini to have commissioned a mappa mundi for the anteroom to his apartments at the

Doge's Palace in Venice in about 1330.[57] His example was soon followed in Siena, where a large circular world map with a disproportionately large depiction of Siena at the centre was created for the audience chamber in the Palazzo Pubblico by Ambrogio Lorenzetti in 1344.[58] The map, which was painted on cloth, was mounted on a wooden disc fixed to the wall in such a way as to allow it to be rotated to ensure that the home city, duchy or kingdom of the guests or envoys being received was always at the top. The circular dents in the plaster left by the rotating world map can still be seen, though the map perished long ago. In the middle of the following century, in about 1448, the Venetian senate commissioned a large world map from Fra Mauro, a Venetian monk in a monastery on the island of Burano, probably for display in the Doge's palace.[59] Scholars continue to argue about whether the map to be seen today in the Biblioteca Marciana in Venice is this map of

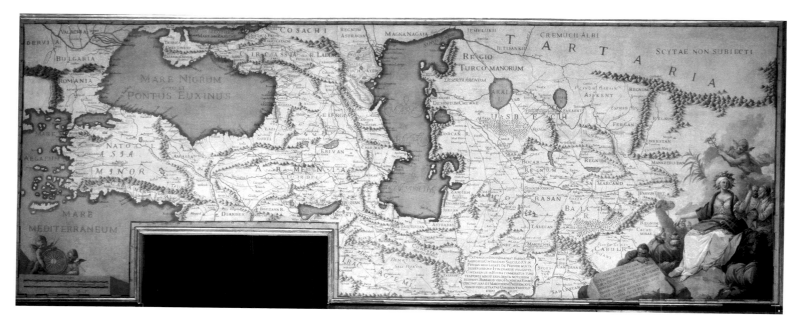

ABOVE: The *Sala dello Scudo*, Doge's Palace, Venice, a reception room leading into the Doge's private apartments. The eighteenth-century painted maps are replacements for maps originally painted in 1540.

RIGHT: The *Sala del Mappamondo*, Palazzo Pubblico, Siena. The space occupied by the map, in front of which the podestà and magistrates stood during receptions, can still be made out.

1448 or a slightly later copy, but it is deservedly one of the best known and splendid of late medieval maps (pp.52–3). Just a dozen or so years later, in 1459, a leading chartmaker Antonio Leonardi created another world map for the Doge's Palace, where it was probably displayed in the antechamber to the principal audience chamber. The practice continued into the sixteenth century. In 1531 Alessandro Zorzi created a world map for the audience chamber itself. In between these dates, and no doubt reflecting the mapmaker's increasing fame, Pope Paul II (1464–71) commissioned Antonio Leonardi to paint a world map for his audience chamber in the Palazzo Venezia in Rome.[60]

In most of these great halls the maps formed part of a wider decorative programme intended to convey messages of secular permanence and power, and often to invoke divine protection for the state. In the *Sala del*

Mappamondo in the Palazzo Pubblico of Siena, for instance, the wall facing the world map is still dominated by Simone Martini's magnificent painting of the Virgin Mary, the city's patron saint and protector, seated in majesty. The room's side walls are covered with depictions, derived from chronicles, of heroic episodes from Sienese history, and the map was surmounted by the bold depiction, probably by Martini, of the mercenary Guidoriccio da Foligno – to whom, along with the Virgin, it was implied, Siena owed its security and prosperity. Similar decoration, often in allegorical form, was to be found in audience chambers elsewhere, sometimes in painted form and sometimes in the form of tapestries.[61] After about 1550 in Protestant Europe, the religious imagery tended to be replaced by portraits of past and present rulers. Even so, and increasingly as time went on, grand manuscript or printed wall maps celebrating famous victories were framed and hung from the walls.

The *Sala del Mappamondo*, created for Cardinal Alessandro Farnese in 1573–4 in his palace at Caprarola near Rome, is one of the most splendid of the surviving cartographic reception rooms (p.8–9). When commissioning the magnificent room, the Cardinal seems to have been influenced by a treatise, published as early as 1510, commenting on the role of cardinal by Paolo Cortesi, who suggested that the reception room of a cardinal's summer palace should contain depictions of engineering works such as irrigation systems (which could be cartographic or quasi-cartographic),

> 'Nor is there less of a learned enjoyment in a picture representing the world, or a depiction of its regions, now known through the audacious circumnavigation of moderns, like that lately done for King Manuel of Portugal in the exploration of India. And the same holds true for paintings exhibiting the remarkable nature of diverse creatures in which the diligence of observation is the more praise-worthy the less familiar are the species portrayed. And in this kind of depiction of curiosities and truths is found that which sharpens the intellect and stocks the mind with learning.'[62]

As late as the 1650s – during what has since been termed the Golden Age of Dutch mapmaking – great dual hemisphere maps of the world were carved into the marble floors of Amsterdam town hall or *Burgerzaal* (p.21).[63] They were located beneath the enormous allegorical sculptures which, taken together, allegorically assert that Amsterdam and its commerce dominated the world.

From the fifteenth century regional maps and maps of a ruler's own territories appeared on public display in the antechambers to the audience chambers. In 1459, or perhaps later,[64] Antonio Leonardi drew large maps of Italy and of the Venetian dominions for the antechamber to the main Venetian audience chamber, providing a replacements in 1497 after the destruction of the original maps by fire in 1484. Leonardi's map of Italy caused something of a sensation. By the early sixteenth century there were copies in the palaces of princes and cardinals of such leading Italian dynasties as the Gonzaga, the Este and the Medici. Likewise, in the 1540s, Henry VIII had a 'platte of Englonde and Scotlande', possibly the so-called 'Cottonian' map of the British Isles now in the British Library (p.102), hanging in the 'Little Galo-rie next the withdrawing chamber' in Hampton Court.[65] In 1591 a 'hall' in Lord Burghley's palatial home at Theobalds, just north of London, contained what was almost certainly an elaborately mounted copy of Christopher Saxton's great wall map of England and Wales, published in 1583.[66]

By the early sixteenth century the chambers of town halls began to be decorated with maps and plans of cities. From the 1530s a wall of the principal reception room in the town hall of Paris was hung with a large gouache (or thick watercolour) map of the city; it was to perish when the town hall was burnt during the troubled period when Paris was ruled by the Commune in 1871, but not before it was photographed.[67] The relatively numerous sur-viving sixteenth-century multi-sheet printed maps of German, Netherlandish and Swiss cities – such as Augsburg, Cologne, Frankfurt, Antwerp, Amster-dam and Zurich – were probably copied from officially commissioned manuscript maps that were once displayed in their town halls. The model for a manuscript bird's-eye view of Great Yarmouth of about 1570, now in the British Library[68], could well have adorned the town hall there from about 1540.

The decoration of the residences of European administrators in the colonies seem to have echoed that to be found at home. In the mid-1640s the walls of the principal reception room of the official residence of the Dutch East India Company's governor on the island of Taiwan, for example, were adorned with images that would not have been out of place in Amsterdam.

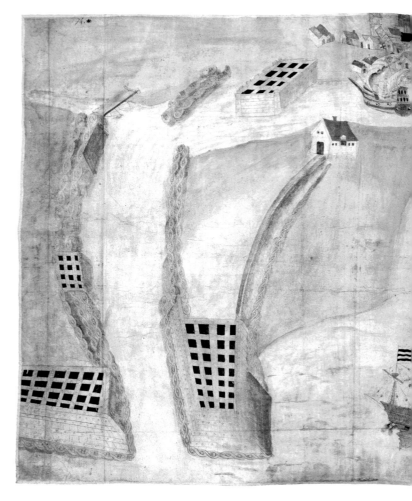

Great Yarmouth in about 1570. The map, containing proposals for the harbour, incorporates what was probably a much earlier town view.
British Library Cotton MS Aug.I.i.74

There were portraits of the princes of Orange, to emphasize their political loy-alties, and religious paintings with a distinctly Calvinist tinge to illustrate their religious ones. Instead of maps of the seven provinces that constituted the Dutch Republic, however, there were three local maps. A map of the whole island of Taiwan would have been a none too subtle reminder that the Dutch had a detailed knowledge of – and could therefore control – the whole island. A map of the principal Dutch fort on the island reinforced the message of authority, as did the map (presumably a bird's-eye view) of a native town.[69]

56 B. Denvir, *From the Middle Ages to the Stuarts: Art, Design and Society before 1689* (London, Longman, 1988), p.48; Daniel Birkholz, *The King's Two maps. Cartography and Culture in Thirteenth-Century England* (London, Routledge, 2004), p.xxv, both quoting the Liberate Rolls.

57 Fiorani, 'Cycles of Painted Maps', p.808, n.20.

58 Marcia Kupfer, 'The Lost Wheel Map of Ambrogio Lorenzetti', *Art Bulletin* 78/2 (June 1996), pp.286–310.

59 There is a voluminous bibliography for the Fra Mauro map. For a recent summary see Peter Barber in P.D.A. Harvey, 2006, pp.42–4, and most recently Angelo Cattaneo, *Fra Mauro's Mappamundi and Fifteenth-Century Venetian Culture* (Brepols, Turnhout, 2009).

60 Fiorani, 'Cycles of Painted Maps', pp.808–9, 813; Schulz, *Cartografia tra scienza e arte*, pp.30–1.

61 For an excellent surviving example, surrounding a wheel of fortune rather than a mappa mundi, see the late thirteenth-century frescoes illustrating Ottone Visconti's exploits in the *Sala di Giusitizia* of the Castle of Angera on Lago Maggiore (Mauro Natale, *The Borromeo Islands and the Angera Fortress* (Milan, Silvana, 2000), pp.143–9.

62 Quoted by Juergen Schulz, 'Maps as Metaphors: Mural Map Cycles of the Italian Renaissance' in David Woodward (ed.), *Art and Cartography. Six Historical Essays*. (Chicago, University of Chicago Press,

1987), pp.117, 228 n.46, from Paolo Cortesi, *De Cardinalatu* (Castel Coresiano, 1510), fols. 54r–v.

63 The present maps were recarved in 1746 by Jacob Martensz (Günter Schilder, *Monumenta Cartographica Neerlandica* vi (Aalphen-aan-den-Rijn, Canaletto, 2000), p.7. The building is now the royal palace.

64 Marica Milanesi points out ('*Antico e moderno nella cartografia umanistica: le grandi carte d'Italia nel Quattrocento'*, *Cartographia Antiqua* XVI–XVII, 2007/8, p,156) that the map of Italy is first *recorded* in 1479.

65 British Library MS Cotton Aug.I.i.9. For a full discussion see Peter Barber, *King Henry's Map of the British Isles, British Library Cotton MS Augustus I.i.9. Commentary* (London, The Folio Society, 2009).

66 Brenchley Rye, p.44.

67 Dérens, pp.25–47.

68 British Library Cotton MS I.i.74.

69 Kees Zandvliet, 'Art and Cartography in the VOC Governor's House in Taiwan', *Mappae Antiquae Liber Amicorum Günter Schilder*, eds. Paula van Gest-van het Schip and Peter van der Krogt (Amsterdam, Hes & De Graaf, 2007), pp.579–94.

Impressing and questioning

In about 1448 the monk Fra Mauro was commissioned by the Venetian Senate to prepare a large world map. Although the Doge is said to have criticized him for not centring the map around an enlarged Venice in the way that its counterpart in Siena had been centred on that city, the space devoted – for the first time on a map – to the feats and observations of the Venetian Marco Polo in Asia occupied the left half the map. This would have amply served to draw the viewer's attention to the achievements of the Serene Republic and its citizens on the world stage. Indeed, so prominent are the texts and illustrations relating to Marco Polo that it was long believed that he was the creator of the map.

The fame of Fra Mauro's map quickly spread, and in 1457 a copy was commissioned by Alfonso V of Portugal. The Portuguese king may have intended to display it in one of his palaces for similarly patriotic purposes. At the top right (south), the Fra Mauro map shows the explorations along the coast of Africa that had been made by the Portuguese since 1415.

This example is an actual-sized copy of the surviving Fra Mauro map, now in the Biblioteca Marciana in Venice. It was commissioned in 1804 by a consortium of influential individuals associated with the East India Company from William Frazer, an artist who specialized in miniatures. The consortium wanted an exact copy because they saw the British as the heirs of the Portuguese in Asia. William Vincent, Dean of Westminster Abbey and one of the protagonists, wrote in a letter of February 1804 to Lord Spencer, a leading politician and book collector, 'as it is the Map copied for the King of Portugal on which the Portuguese discoveries depended, a copy of it may be worth bringing to England' (British Library Add. MS 76092).

Fra Mauro's map was epoch-making in several ways. Much of its content is derived from the Bible and the same classical writers whose influence is found in medieval world maps, while the coastlines are mainly copied from contemporary Mediterranean sea charts. Fra Mauro quotes from the writings of Ptolemy and the location of regions and places in the interiors of India and Africa are derived from his co-ordinates. The finished map brings together the three main strands of fifteenth-century mapmaking: the biblical, the maritime and the Ptolemaic. The overall appearance is visually pleasing and it has been suggested that Leonardo Bellini painted the scene of the Garden of Eden (shown left).

Fra Mauro was no mindless compiler, however, and the texts on the map reflect his questioning attitude to all three historical traditions. Such an approach was completely novel, and was partly based on his discussions with merchants and seamen in Venice.

Fra Mauro could find no evidence for the existence of the Garden of Eden on Earth, and so he daringly places it just outside the world outline in a corner of the map. For similar reasons he shows none of the monstrous peoples to be found on earlier maps, and his text emphasizes that many of the outlines derived from Ptolemy's co-ordinates were incorrect. Contrary to Ptolemy he does not show the Indian Ocean as an inland sea, and he anticipates the rounding of the Cape of Good Hope by about forty years. In its questioning attitude the map acts as a bridge between medieval and modern mapping.

This copy is very close to the original except that the lettering has been modernized and contractions expanded, the two insets at the bottom have been reversed and the landscape outside the Garden of Eden slightly modified.

Cattaneo (2009).

The Fra Mauro World Map copied by William Frazer, London and Venice, 1804
Manuscript on vellum, 239 x 229 cm
British Library Add. MS 11267

A free imperial city

Jörg Seld (engraved by Hans Weiditz), *Sacri Romani Imperii Civitas Augusta Vindelicorum*, Augsburg, Grym & Wirsung, 1521

Woodcut on eight sheets, 80 x 191 cm

British Library Maps *30415. (6.)

This is the first printed European town plan based on mathematical measurement to be produced north of the Alps, and only the second such map after De' Barbari's bird's-eye view of Venice (pp.30–1). It celebrates Augsburg in a manner that encapsulates the essence of German Renaissance humanism: a veneration for the classical past, great civic pride about the existing city and the application of reason, through mathematics, for practical, patriotic purposes.

The mapmaker Jörg Seld, Augsburg's leading goldsmith, together with its publishers Sigmund Grym, the town physician, and Markus Wirsung, an apothecary, formed part of the City's mercantile-humanist elite. Grym was married to a member of the Welser family, a banking dynasty who with another banking dynasty, the Fuggers, dominated the city. He was also connected to Konrad Peutinger, a merchant and an early owner of the 'Peutinger Table' (p.13).

Seld had surveyed the city in 1514–16 and may have completed a manuscript map for display in the town hall. The manuscript was only turned into a woodcut and published in 1521, however, probably in connection with celebrations for the imperial grant to Augsburg, in that year, of the right to mint its own coins. This in effect constituted official recognition of Augsburg's independence under its suzerain lord, the Emperor Charles V.

The large text panel at the right of the woodcut summarizes the story of the imperial protection that the city had enjoyed since the time of its founder, the Emperor Augustus, and invokes the continuing protection of God, Charles V, and the Holy Roman Empire as a whole. The imperial and Habsburg coats of arms at the top left reinforce this message. The panel on the left states that the print is intended for those homesick for Augsburg, and expresses the hope that the image will satisfy the hungry eyes of those who, attracted by its fame, are anxious to know the city better.

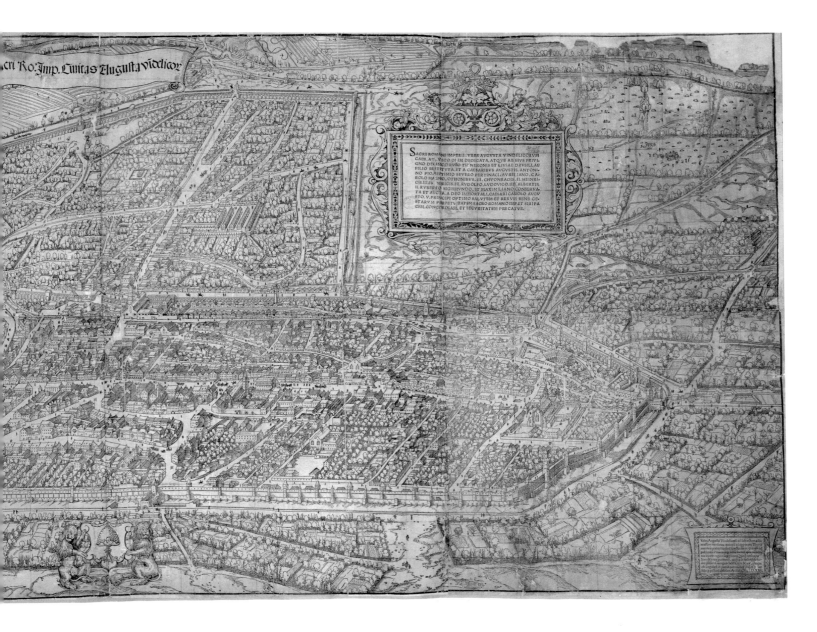

The publishers of Seld's map had hitherto specialized in publishing texts by classical and early Italian writers, and the cutter of the wood blocks, Hans Weiditz, has been identified as the artist who provided high-quality illustrations for their books. Thus although, as a map, it was atypical of Grym and Wirsung's press, its purpose and context were entirely in line with the publishers' interests in a city that regarded itself as a new Rome north of the Alps.

Side-by-side with patriotic pride in its past went a fierce pride in the existing city and the intellectual and practical abilities of its inhabitants. The text at the bottom right emphasizes that the image of the city was based on the exact measurements made by Seld. He shows the Fuggerei, a complex of almshouses to the north of the city walls, which opened in 1519 and still survives. Elsewhere the elevations of the principal secular and religious buildings can be seen with Augsburg's citizens, in silhouette,

going about their business (including a funeral taking place at the top left). The identification of many of the buildings has been added in a contemporary hand and the print has a pencilled grid demonstrating that it was squared for copying, perhaps in the early nineteenth century when a lithograph copy was published.

The city of Augsburg owns the only other example of the original state, or first printing of this map. It is richly coloured and may have been the official presentation copy intended for display in the city hall. Copies of the map were sold throughout Europe, however, and its influence can clearly be discerned in William Cunningham's bird's-eye view of Norwich of 1559, the earliest surviving English printed town plan, and in the woodcut 'Agas' map of London that first appeared in the 1560s.

Geisberg (1974), iv, pp.1493–7; Schulz (1978), pp.468–70.

Lands, friends and loyalty

Details from the Sheldon map of Oxfordshire, including London (right)

Map tapestries can be regarded as the north European equivalent of frescoes: each was well adapted to the different climates of their respective regions. Tapestries and frescoes were prestigious, luxury products that were colourful, covered large areas of wall and proved extremely expensive to produce. From at least 1520 map tapestries were being commissioned by, and for, emperors and kings.

By about 1590 tapestries were also being commissioned by a wealthy English landowner, Ralph Sheldon, to adorn the four walls of the principal chamber of his house at Weston in Long Compton, Warwickshire. There were four tapestries which collectively showed the Midland counties of Gloucester, Warwick, Worcester and Oxford where Sheldon had lands. The tapestries were no mere display of wealth, however. The emphasis on the lands and home of the Sheldon family and their Catholic friends and relatives, the royal arms in the top left corner, a panel containing appropriate descriptive text from *Britannia* (the great and recently published topographical work by William Camden), in the top right corner, along with the Sheldon arms at the lower right combined to send a powerful message which would have easily been understood at the time. The Sheldon family, it was made clear, were powerful but, though Catholic, they were patriotic and loyal supporters of the Protestant Queen Elizabeth, even possessing close links to her favourite, Robert, Earl of Leicester whose houses were depicted recognizably. In the years after the Armada, which were so perilous for English Catholics, these were points that needed making.

The Sheldon family homes, and probably their tapestries, suffered serious damage during the Civil War, and it was shortly after the Restoration in 1660 that Ralph's grandson, also Ralph (1623–84), commissioned replacements for two of the tapestries, probably from weavers in the recently established factory at Mortlake. An antiquary by inclination, this Ralph had the map content copied. They were not, however, precise copies, featuring some updatings and minor additions and a new and more fashionable woven frame. Though persecution of Catholics did continue – Ralph himself was arrested during the so-called Popish Plot scare of 1678 – it had diminished, and perhaps there was no longer the need for the second Ralph to proclaim his loyalty to the Crown. For whatever reason, the royal arms and the panels with text by Camden which had given the originals such force are omitted. Nevertheless the tapestry continues to send powerful messages about the family's continuing power and wealth. The arms are those of Ralph and his wife, who died in 1663.

This tapestry, which is centred on Oxfordshire, is one of the replacements. Because of the shape of the tapestry its coverage extends as far as London (with a domed St Paul's Cathedral) at the bottom right, though the boundaries of Oxfordshire are highlighted in red and each of the other counties has a differently coloured background. The mapping is based on Christopher Saxton's surveys of the 1570s, but other sources, apparently including direct observation, have been used. Most towns are individualized, as are the homes of Sheldon, his relatives, friends and associates, including Osterley and Syon House near London, but several other large houses are omitted. A further touch of local colour is added by the depiction of the White Horse that gives its name to the Wiltshire vale – though it looks more like a cart horse and has been turned around to face south.

Turner (2007), pp.67–72; Turner (April 2002), pp.293–313; Turner (June 2003), pp.39–49.

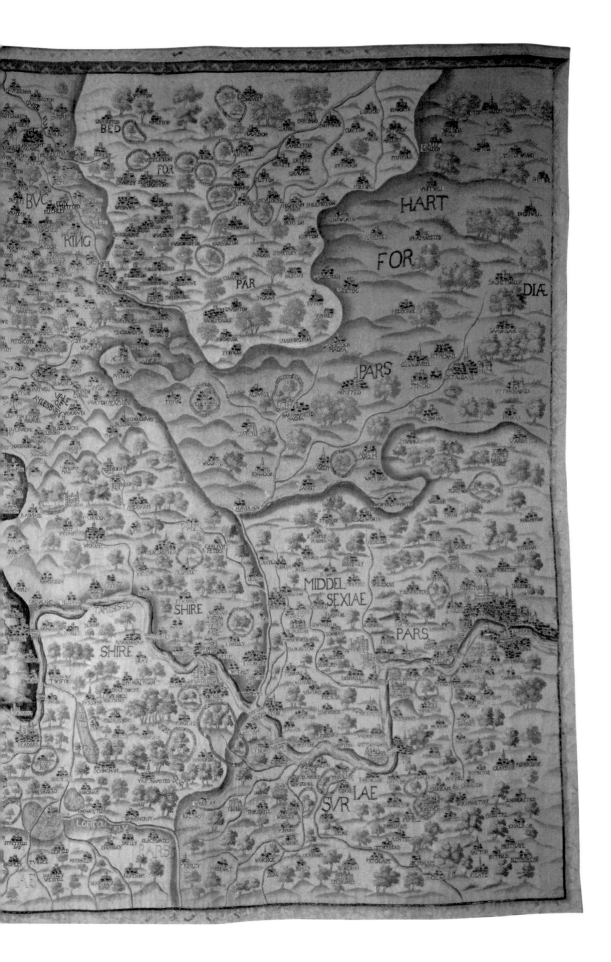

Map of Oxfordshire, Mortlake, *c.*1663
Tapestry, 445 x 600 cm
The Sheldon Trust

Spheres of influence

During the final decade of the reign of Elizabeth I, England had emerged with enhanced importance on the world stage. In common with the rest of Europe the country was experiencing a period of considerable economic prosperity, and its increased importance as a maritime power, underlined by the defeat of the Spanish Armada in 1588, was as relevant for trade as for national security. There were repeated English voyages to the Americas, often combined with a search for new commercial routes to the riches of the East Indies and China, particularly the elusive North-West Passage between North America and the Arctic. Such voyages were aided by scientific advances in the form of improved navigational instruments, charts and globes. Yet these expeditions, and the imperial-minded individuals who financed and led them, relied on publicity, support, and above all on open displays of favour from the queen.

The terrestrial and celestial globes made in London in 1592 by the mathematician Emery Molyneux (d.1598/9) attracted great attention when they first appeared. They were praised as the 'fairest paire of globes in England,' and motivated a number of treatises by scientists on their correct interpretation and use. They were the largest printed globes to have been produced up to that time. Financed by the merchant Sir William Sanderson (1547/8–1638), sets of the globes were presented to Elizabeth on two separate occasions in 1592 and 1594.

Molyneux's globes are significant for the expression of English imperial power through the marking on the terrestrial sphere of the circumnavigation routes of Sir Francis Drake and Thomas Cavendish, among others, and the naming of English colonies in Virginia and Guiana. These features, and the placing of Elizabeth's arms over North America, served to symbolize the promoters' ambitions for their country, reinforced by the dedication requesting the queen's support of naval expansion. Sanderson's patronage of the globes was clearly intended to gain, through flattery, the support of the queen for his further ventures; he had previously financed voyages by Raleigh and by Frobisher, with the aim of reaping the financial rewards and commercial links which might have followed. Evidently Sanderson hoped to win personal favour with his queen through glorifying her and associating her with imperial rule over North America.

Emery Molyneux, *Terrestrial and Celestial Globes*, London, 1592/1603

Mixed media (papier mâché, plaster and printed paper) with wooden stands and brass surrounds, diameters 63.5 cm

By kind permission of the Masters of the Bench of the Honourable Society of the Middle Temple, London

Molyneux's globes represent the pinnacle of scientific achievement in Tudor England, yet their creation is interlinked with Dutch globe production. The gores, or map images printed on paper forming the outermost layer of the globes, were engraved by Jodocus Hondius during his time in England from 1584 to 1593, and the celestial gores copied from a globe of 1589 by Florent van Langeren. The terrestrial globe was updated to show discoveries to 1603. This pair, two of only five surviving globes, are owned by Middle Temple Library in London with whom they have shared a long and close history.

Wallis (1951), pp.275–90; Wallis (1955), pp.304–11; Wallis (1962), pp.267–79.

An explanation of power

John Speed's map of England and Wales, an earlier manuscript version of which is recorded as having been displayed in the gallery of James I in Whitehall Palace, emphasizes the validity of the monarch's claim to the throne of England. The power it implies is of the highest level, and its message confirms the security of England under a Scottish king. These would have been clear to all who looked upon it in his presence.

The accession of James I (1566–1625) to the throne of England in 1603 combined for the first time the thrones of England and Scotland, yet the legitimacy of James's claim was by no means accepted by all. In its early years, the power of the Stuart monarchy depended on its claim being asserted with skill and conviction, and visual propaganda such as maps were an intrinsic part of the publicity drive. The genealogical tree, on the right of the map, shows the line of English monarchs stretching from William the Conqueror, with James placed authoritatively at the top. Such a clear lineage served to 'correct' any prior misgivings.

More than simply validating James's rule, Speed goes on to explain the instruments through which this rule is exercised. The king's power runs through the reins of government represented by the arms of secular noblemen and the bishops below. Figures in each corner symbolize the ruling estates of the kingdom: the monarch, the Lords spiritual and temporal, and the Commons. The image of the king at the head of a solid and well-organized system of government, including everybody under the new term of 'the United Kingdom', would have spoken to all who had experienced political turmoil in the past, just as many references to previous wars and battles would have reminded people of their uncertain past, while assuring them of a safe future.

In popularizing the security of government under the new and legitimate ruler, Speed's map would have been complemented by other maps in the gallery which served as reminders of previous eras of English history: tapestries, globes and maps from the collections of the Tudor monarchs. The historian and compiler John Speed (c.1552–1629) is thought to have produced this four-sheet map hurriedly from his earlier manuscript map, in order to assert copyright against a pirated edition produced by Hans Woutneel. Two examples of the map are known, the British Library's version missing the top-right sheet. A replacement photograph of this has been taken from the example in the Bibliothèque nationale de France in Paris.

Barber (2007), p.1659; Shirley (1991), no.261, pp.106–7; Schilder and Wallis (1989), pp.22–6.

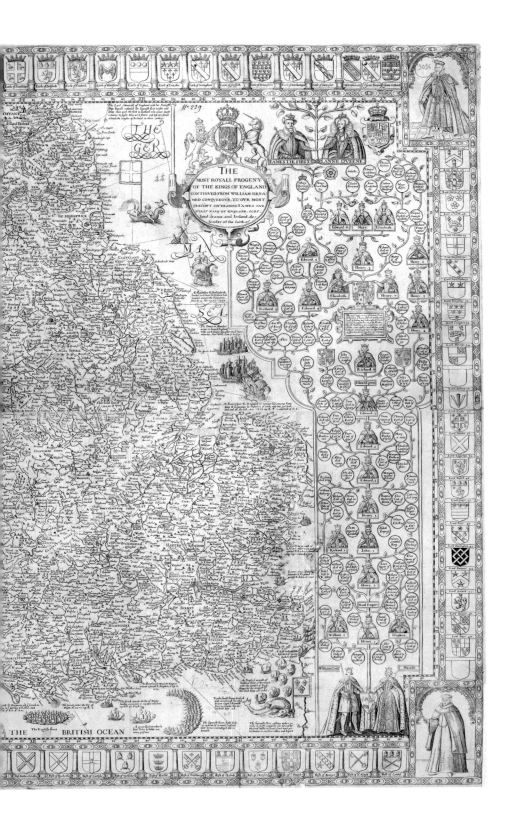

John Speed, [*Map of England, Wales and Ireland*],
London, 1603–4

Copperplate engraving on four sheets, 83 x 110 cm

British Library Maps 188.t.1.(1.)

A nation with a map

This map of the United States of 1784 employs the same language of national pride as the great European survey maps of the sixteenth and seventeenth centuries. It consequently shares more of an affinity with the arena of overt power than with the comparatively subtle nuances of the gallery. Unlike these earlier maps, such as Philipp Apian's 1568 map of Bavaria or Lubin's Pomerania of 1618, both the result of original surveys, its geography is largely derivative. Yet this is irrelevant to the significance of the map, claimed by its author Abel Buell (1742– 1825) to be 'the first ever published, engraved and finished by one man, and an American'. Clearly such a great accolade would have favourably influenced sales of the map to the sort of 'patriotic gentleman' identified in Buell's advertisement.

Although the commercial motives behind such patriotic language may be clear, the powerful display of national identity is best appreciated in the context of a nation still in the early years of its existence. The Declaration of American Independence had been signed in 1776, but a further seven years would ensue before the cessation of hostilities with the British and the signing of the peace treaty in 1783, one year before the publication of Buell's map. The symbolism of the map acts accordingly, claiming geography – and, by implication, the land – from its previous colonial ruler through the adoption of the meridian of the United States capital of Philadelphia, rather than the prime meridians of the Azores, London or Paris which were then commonly used. The future direction of American policy is also hinted at in the western expansion of the United States' dominion in strips across the entire continent (stopping just short of the Mississippi River), oblivious to the existing rights and claims of the Spanish, French and Native Americans. At the same time, classical references underline the endurance of Thirteen States, with the figure of Liberty in one of her earliest American guises, and the date of independence written as if carved into stone. In line with Renaissance ideals, this map is referencing the old in order to reinforce the new.

To say that Buell's map is truly national might be overstepping the mark. Local as well as national loyalties existed in this vast new country, stretching from Maine in the north to Georgia in the south. The map, in fact, shows a particular bias towards the New England state of Connecticut through its dedication to the state governor and the arms of the state below the Stars and Stripes. Buell was a silversmith from New Haven and not known for making maps. It is possible that he decided to stray into the map trade by seeing the opportunity to capitalize on the patriotism created by the defeat of the British a year previously. What Buell's map signifies is not so much a nation in need of a map as a mapmaker in possession of a nation.

Wroth (1958).

Abel Buell, *A New and Correct Map of the United States of North America…*, New Haven, 1784
Copperplate engraving on four sheets, 160 x 127 cm
British Library Maps *71490.(150.)

Pomerania matters!

The rulers of small states often sought to enhance their international standing – and to distract attention from their lack of political might – by focusing upon their cultural and scientific accomplishments. Such attributes neatly combined with patriotism when they commissioned sophisticated maps of their lands.

This was the case with Duke Philip II of Pomerania, a land of coastal towns and a sandy hinterland that stretched along the southern shores of the Baltic Sea. This strategic position led its powerful neighbours Sweden and Brandenburg to cast envious eyes on it. But when Philip commissioned Erhard Lubin, or Lübben, a professor of poetry and theology at Rostock University, to undertake a detailed survey of his dominions 'for the special fame of our duchy and lands' in 1610, the prospects for Pomerania looked more rosy than they had for many years. The succession seemed assured since Philip had four brothers and there was a likelihood that following the death of a childless cousin, Philip Julius of Wolgast, all of the Pomeranian dominions would be re-united under him.

Lubin carried out his survey between 1611–12, with the duke's active participation. He used tried and tested methods and instruments rather than following the latest recommendations of theorists, revising the map in 1617. At an approximate scale of 1:200,000 (about 2 kilometres to the centimetre or 3 miles to the inch), it shows even the smallest hamlet. In 1614 Duke Philip commissioned Hans Wolfart (Wolffrodt), an artist originally from Antwerp, to depict the towns, and he ordered the nobility and cities to provide examples of their arms. The resulting views and map were sent for engraving to Amsterdam, the European centre for such work. The finished work was engraved in twelve sheets by Nicolaes van Geilkercken and published in 1618.

Lubin's map is an epitome of the country, with a portrait of Philip surmounting those of his younger brothers and his cousin Philip Julius, the ruling Duke of Wolgast. At the top there are two family trees, one showing the descent of the early dukes of Rügen from 938 to 1325 and the other, illustrated with portraits, showing the descent of the currently ruling family from 1107. No less than forty-nine Pomeranian towns and the arms of 353 noble families are depicted. At the foot of the map Lubin composed an extensive written description of the duchy, explaining its history, geography, geology and economy – including a list of the fish to be found in its waters. He portrayed himself, surrounded by his surveying instruments, above the scale bar at the bottom-right of the left-hand text panel. Very few copies of the map were printed at the time as it proved impossible to find paper of sufficient quality in Pomerania to take impressions of the copperplates.

Duke Philip died in February 1618, before he could be presented with the finished map, and in the following years all hopes for the salvation of his country gradually evaporated. Pomerania was ravaged by repeated invasions during the Thirty Years' War, exposing the powerlessness of his brother and successor Bogislav XIV. Following Bogislav's death in 1637 the royal line became extinct and the country was partitioned, with Sweden taking Stettin and the western lands, and Brandenburg taking eastern Pomerania.

Despite the low print run, Lubin's map provided a model for later maps of the area. The original copperplates were lost until their chance discovery in Stralsund in 1756 and a few new copies, including this example, were printed in Hamburg in 1758.

Haas (1980).

OVERLEAF:
Erhard Lubin (engraved by Nicolaes van Geilkercken), *Nova illustrissimi Principatus Pomeraniae Descriptio*, Amsterdam, 1618
Copperplate engraving on twelve sheets, 133 x 218 cm
British Library Maps *29755.(8.)

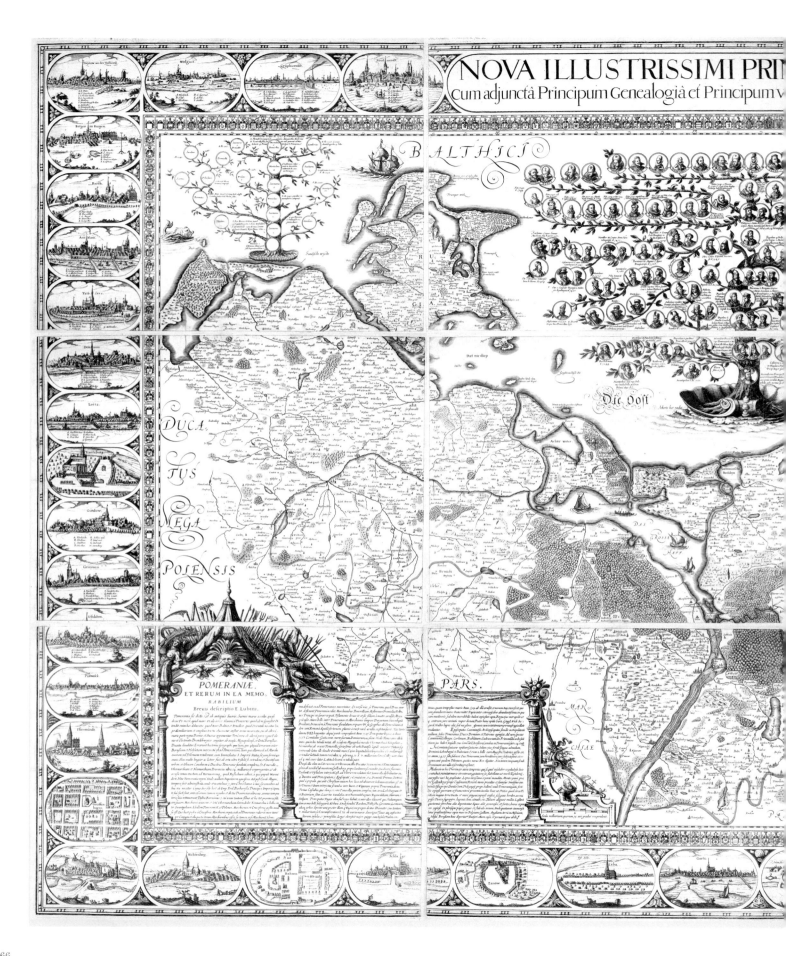

NOVA ILLUSTRISSIMI PRIN
cum adjunctâ Principum Genealogiâ et Principum v

BALTHICI

Die Oost

DUCA
TUS
MEGA
POLENSIS

POMERANIAE,
ET RERUM IN EA MEMO,
RABILIUM
Brevis descriptio E. Lubini.

PARS.

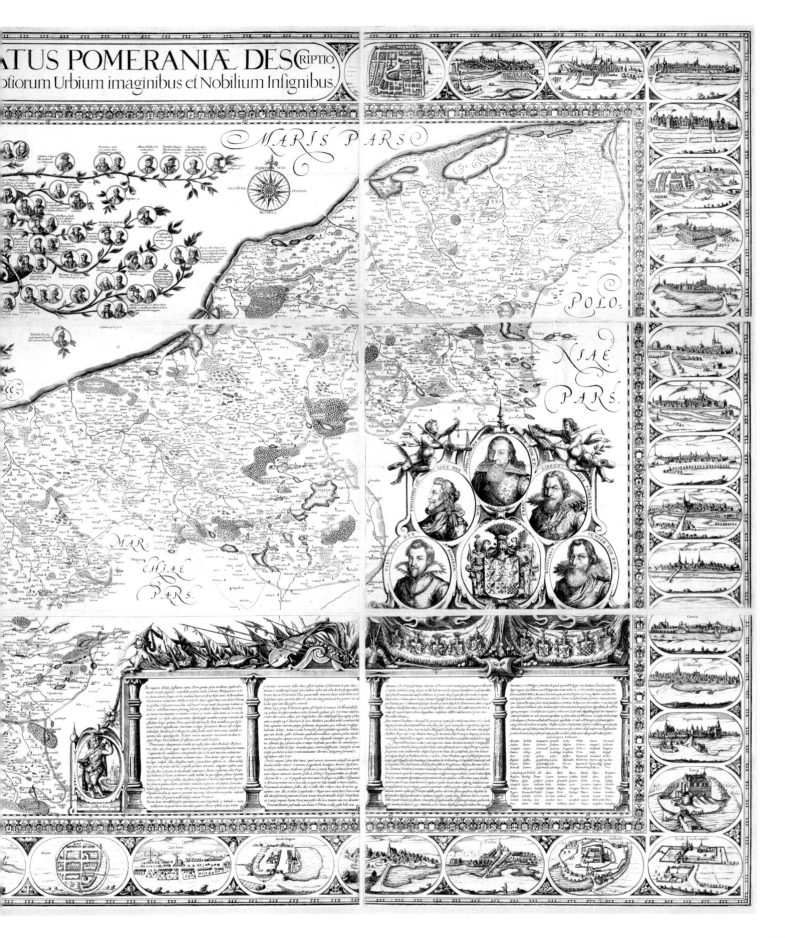

A famous victory

The Infanta Isabella, regent of the southern Netherlands on Spain's behalf, may have sent this very map to one of her fellow European rulers in the spring of 1628, together with an explanatory booklet by Alonso Ladrón de Guevara with its title page designed by Rubens. The map commemorates what she considered her greatest military triumph. The absence of a title at the top and of surrounding letterpress text show it is the very first state of the map before changes insisted on by the mapmaker when ordering copies for wider circulation. Its stained state suggests that it was varnished for display in the same way that the oil paintings hanging near it in the palace were (and paintings still are) and in the mistaken belief that the map would thereby be protected.

Breda, a Dutch enclave in the southern Netherlands, commanded the route to Utrecht and Amsterdam in the Dutch Republic. Following the renewal of hostilities with the Dutch in 1621, the city became an obvious target for the Regent. Between August 1624 and June 1625 a siege led by Spain's most brilliant general, Spinola, succeeded in starving the Dutch garrison into submission. Within weeks the Infanta, acting through Spinola, had obtained the Duke of Lorraine's permission for his court artist, Jacques Callot, to prepare a map commemorating the successful siege.

Callot had already made his name as a supremely talented etcher in Florence before returning in 1621 to his home city of Nancy in Lorraine. He visited Breda at least three times, actively surveying the area, and took two years to complete the map. Its content was much influenced by the published, illustrated account of the siege by Hermannus Hugo, one of Spinola's chaplains, and by printed Dutch maps and news accounts. The angle of vision moves from bird's-eye view at the bottom to plan (vertical and mathematically calculated) and finally to elevation (the normal, unelevated viewpoint with a horizon) at the top. The map, in the tradition of history paintings stretching back to 1500, combines numerous episodes from the siege in one image. At the bottom-right Isabella can twice be seen in her state carriage approaching Breda for her triumphant entrance on 11 June, while the top of the map features the Dutch garrison retreating to Gertruydenberg. At the top left are the arms of Spain and at the top right the 'lozenge' containing the arms of the Infanta. Spinola's arms are at the bottom centre, in recognition of his feat of arms and his sponsorship of the map. At the bottom left Callot has shown himself at work, being instructed by one of the Florentine military engineers whom Spinola had recommended.

At the same time as functioning as a public celebration the map, considered by many to be one of Callot's masterpieces, was very much a personal statement. The realities of seventeenth-century warfare are illustrated throughout the map, with repeated vignettes of pillage, torture and hangings. These may derive from Hugo's account of the siege, but Callot not only chose to depict them here – he also repeated them a few years later in his famous series of etchings on the miseries of war.

There is much in this map that would qualify it as a work of art. In 1633–4 the Spanish court painter Diego Velázquez incorporated several details from Callot's map, including the plan of Breda, into the background of his famous painting of the Surrender of Breda, commissioned by Philip IV for a gallery of paintings of military triumphs for his palace of Buen Retiro in Madrid. A detailed comparison between Callot's map and Velázquez's painting demonstrates how close painting and mapmaking could still be in the early seventeenth century.

Zurawski (1988), pp.621–39; Lieure (1927), ii, no.593.

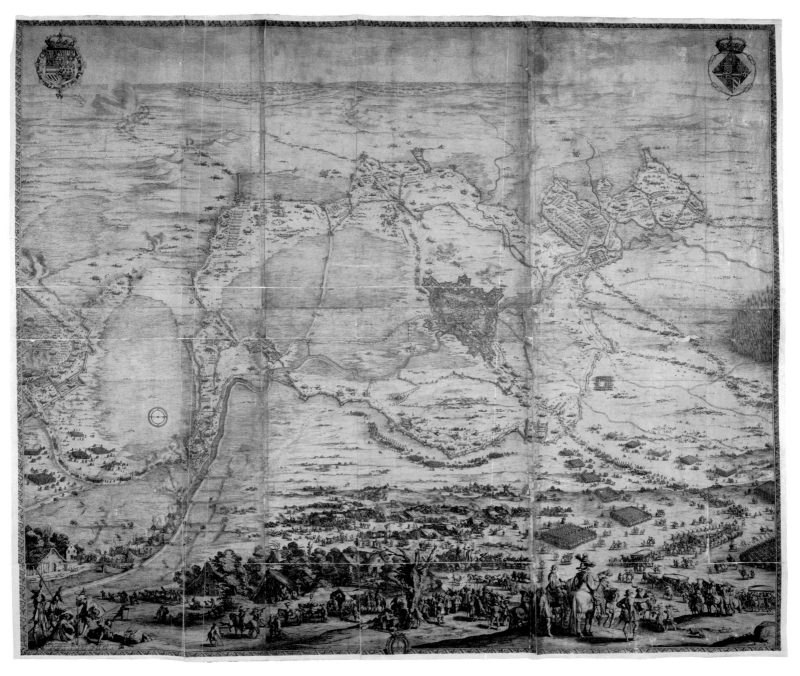

Jacques Callot, *The Siege of Breda*, Antwerp,
Balthazar Moretus, 1627/8

Etching on six sheets, 119 x 145 cm

British Library Maps * 32760. (1.)

Glorifying the kingdom not the king

This map of Bohemia is, arguably, the greatest cartographic achievement of the Central European Baroque. It is the work of an outstandingly competent military engineer, Johann Christoph Müller of Nuremberg (1673–1721). The 25-sheet map, at a scale of approximately 1:132,000 (about 1.3 kilometres to the centimetre or 2 miles to the inch), contains the astronomically determined locations of no less than 12,500 settlements and natural features. It has fifty signs for human settlements, physical relief, political and administrative boundaries, military installations, agriculture, industry, commerce, communications, religion and history. Unsurprisingly, it was to remain the standard printed map of the region for the next century. At the same time, and aided by the need to fill the corners around the circular shape of Bohemia, it is an outstanding work of art. The decoration is the work of Wenzel (Vaclav) Lorenz Reiner (1683–1743), a leading artist of the Czech Baroque, whose frescoes adorn numerous churches in and around Prague. It was engraved by one of the most accomplished German engravers of the time, Johann Daniel Hertz (1693–1754).

In 1708 the Austrian War Council in Vienna commissioned Müller, at the latter's suggestion, to survey in detail the kingdom of Bohemia, the marquisate of Moravia and the duchy of Silesia which constituted the political unit commonly known as the Bohemian crown lands. The kingdom of Bohemia was the wealthiest and most productive region of the Habsburg empire, producing from direct taxation more than half its revenues. It was, though, particularly exposed to invasion, suffered peasant unrest and had a large Jewish population, all of which were of considerable interest to the authorities. Müller surveyed Bohemia between 1712–18. In 1720 he presented the emperor and estates of Bohemia with very large-scale manuscript maps of the individual districts of the province and a map

St Wenceslas offering Prague to the Virgin

of the kingdom as a whole, at a much-reduced scale. At the same time a working copy was sent to Augsburg for engraving by Michael Kauffer, the figurative work being subcontracted to Hertz in Nuremberg. The map that we see was completed two years later.

Because of the chronic poverty of Habsburg central government, the main cost of the survey was borne by the Bohemian estates (or parliament), dominated by its wealthy and powerful nobility. At the time these aristocrats were growing increasingly resentful of the emperor's neglect, Charles proving much more interested in Spain, Italy, Germany and Hungary. Such resentment may be detected in the splendid decoration around the map. Unlike almost all the major decorative sequences commissioned during Charles VI's reign, there is no mention whatsoever of the emperor (see p.151).

At the top-left, and in line with Habsburg-sponsored strident Counter-Reformation Catholicism, can be seen the patron saint and first Christian king of Bohemia, St Wenceslas (the 'Good King Wenceslas' of

the carol, who was assassinated in Stará Boleslav in 923). He is shown recommending the city of Prague (seen below) to the protection of the Madonna of Stará Boleslav (Altbunzlau). In the right corner, male and female figures, representing Bohemia's rivers, surround the crowned, double-tailed heraldic Bohemian lion, while to the left a landscape, including horse and cattle breeding, hints at the kingdom's fertility. The theme is continued at the bottom right with references to agriculture, wine and beer making and hunting while at the left, around the reference table, Czech industry is represented by scenes of mining, glass-blowing and stone masonry.

Copies of Müller's map adorned the Prague and Vienna palaces of Bohemia's proud Catholic nobility. Today, several revolutions later and long after the collapse of the Habsburg Monarchy in 1918, one embellishes the official residence of the ambassador to London of the Czech Republic.

Kucharl (1961), pp.24–31; Dörflinger (1989), pp.70–5, 318–20; Evans (2006), pp.75–98.

Allegory of Bohemia's fertility

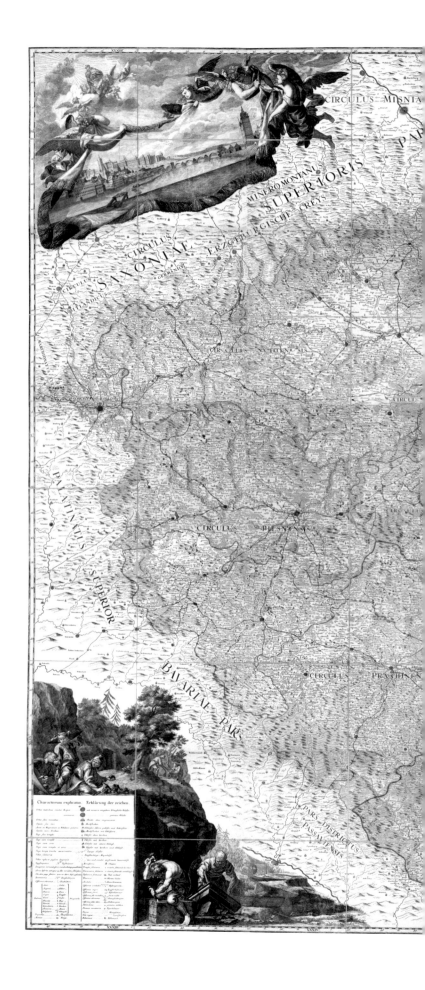

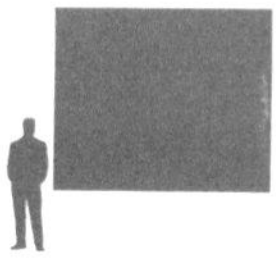

LU-SATIÆ SUPERIORIS PARS

DUCATUS TATARIENSIS

DUCATUS

DUCATUS SWIDNICENSIS

SILESIÆ PARS

DUCAT. MON-STERBERGEN-SIS

MERICENSIS

CIRCULUS BOLESLAVIENSIS

CIRCULUS REGINOHRADECENSIS

PARS DUCATUS KROTKAVI ENSIS

COMITATUS GLACENSIS

CIRCULUS KAURZIMENSIS

CIRCULUS OLOMUC.

CIRCULUS CHRUDIMENSIS

PARS

CIRCULUS CZASLAVIENSIS

MARCHIONATUS

MORAVIÆ PARS

SCIRCUL BRUNNENS

CIRCULUS BECHINENSIS

CIRCUL IGLAVIENSIS

AUSTRIÆ PARS

HIDUCATUS

MAPPA
GEOGRAPHICA
REGNI
BOHEMIAE
IN DUODECIM CIRCULOS DIVISÆ
CUM COMITATU GLACENSI
ET DISTRICTU EGERANO
ADIUNCTIS CIRCUMIACENTIUM REGIONUM PARTIBUS CONTERMINIS
ex accurata totius Regni perlustratione
et geometrica dimensione
OMNIBUS, UT PAR EST NUMERIS ABSOLUTA
et
ad usum communibus
nec non amœnæ et singula delineatis cognoscenda
XXV SECTIONIBUS
exhibita
à
Joh. Christoph. Müller, S.C.M.Capitan. et Ingen.
A.C. MDCCXX

Johann Christoph Müller, *Mappa geographica regni Bohemiae in duodecim circulos divisae … MDCCXX*, Prague, 1722

Copperplate engraving on twenty-five sheets, 240 x 282 cm

British Library Maps 186.k.2./ British Library 118.a.3-7

73

An impression of power

We do not know the precise circumstances that led Michelangelo Blasco to prepare this manuscript map of northern Italy for presentation to the Holy Roman Emperor, Francis I, and his much more powerful wife, Maria Theresa of Austria, in about 1760. Blasco was a leading military engineer who had built forts in Sicily for the Bourbon king of Naples in the late 1730s. In 1752 he had been responsible for delineating the southern borders of Brazil and Spanish America. Since, in order to create the map, Blasco would have needed access to confidential manuscript maps, conserved only in Milan and Vienna, it can be assumed that the commission was official, was meant for display at court and was intended to impress.

And impress it certainly does – not only through size and dramatic decoration, but also through content. As the map proclaims, it is a reduced version of a survey of Austrian Lombardy, the so-called *Censimento*. This had been undertaken to increase tax revenues between 1720 and 1723 under the leadership of Johann Jakob Marinoni (1676–1755), court mathematician and astronomer, and co-founder in 1718 of the imperial academy for military engineers in Vienna. As text on Blasco's map also states, the

surveyors had used the 'Pretorian' plane table ('Tavoletta Pretoriana'), invented by Marinoni. The instrument made possible the relatively speedy creation of accurate maps at a scale of 1: 2000 (500 centimetres to the kilometre or 30 inches to the mile), enabling every single field and building to be depicted. It may be worth bearing in mind that it was to be another 120 years before the British Ordnance Survey began mapping regions of Great Britain at that scale: although Britain had the requisite technology, the landed aristocracy which dominated its parliament until the Reform Act of 1832 would not allow detailed mapping of their estates, regarding it as 'snooping' by the central government. From these very detailed surveys parish maps were produced at 1:8000 (just over 12 cm to the kilometre or about 8 inches to the mile) and a military, topographical map at 1:72,000 (about one centimetre to three-quarters of a kilometre or about an inch to a mile). The Austrian surveyors used the mapping process not only to fix Austria's international boundaries but also, less successfully, to extend imperial authority throughout northern Italy by arguing that the emperor must have a right to any land that his surveyors could map. The maps

Michelangelo Blasco, '*Stato di Milano e sue provincie tirate dal Censimento o sia perticato di esso stato ...*', Milan or Vienna, *c*.1760
Manuscript on paper, 130 x 230 cm
British Library Add. MS 68933

remained in manuscript, available only to ministers and officialdom on a 'need-to-know' basis.

In order to produce a wall map from the original *Censimento*, Blasco's map is so much reduced, at a scale of about 1:132.300 (1.3 km to the centimetre or about two miles to the inch), that all potentially sensitive information has been omitted. Nevertheless, what is left is a far more accurate map of Lombardy than was otherwise available. As Blasco's inscriptions on the map state, he has intentionally distinguished cartographically between the advanced style in which the Habsburg lands are depicted, with the layout of even moderately sized settlements being shown in plan, and the more traditional manner in which the lands of neighbouring rulers are shown, where the locations of settlements are indicated by symbolic towers. Any onlooker would surely have been impressed by the sophistication of the Habsburg mapmakers, and by implication of their administration, in contrast to the cartographic backwardness of the neighbouring Italian states, such as Savoy-Sardinia and Parma. The grandeur of the Piranesi-style decoration would have further proclaimed the military and political power of the Austrians as compared to their neighbours, and hinted at the legitimacy of the imperial couple's rule in Italy by recalling the Roman origins of the Holy Roman Empire.

Blasco's image of northern Italy did not fully correspond to reality. The Sardinians, in particular, had produced some equally accurate and detailed maps of their territories from 1718, though they had remained in manuscript and were confidential. Moreover, as any informed contemporary would have recalled, the Austrians had come very close to being driven out of Italy by Bourbon forces led by Blasco's former master, Don Carlos, in the 1730s. However, the Habsburg authorities must have hoped that visitors who saw the masterful cartographic image would take it for the continuing reality.

Kain and Baigent (1992), pp.181–7.

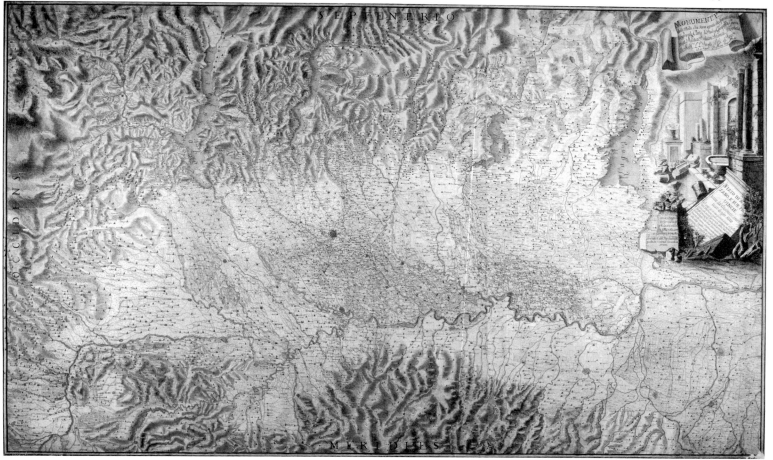

The diminutive Psalter map (p.78), drawn in a workshop in Westminster in the early 1260s and now in the British Library, is probably a reduced copy of a very large world map commissioned by Henry III in 1236 for the bedchamber in his palace of Westminster. Of possible surprise to twenty-first century readers, the bedchamber was the most important space in the palace after the Great Hall. In some senses it was more important, since it gave visitors direct and close access to the person of the king, the embodiment of power.[70] The bedchamber was usually a large room within which the structure of the great bed, or bed of state, conveyed connotations of power. The bed consisted of an inner bed on which the king (sometimes) slept and an outer bed which could be used as a throne. The bed's canopy was identical to the canopy of state under which the ruler appeared on ceremonial occasions. As well as being used as a reception room, Henry III's bedchamber was used for meetings of parliament and for great ceremonial occasions, as in 1270 when the king of Scotland paid homage to Henry.

The bedchamber at Westminster came to be known as the Painted Chamber because of its splendid decoration – religious, philosophical and encyclopaedic in nature. Before a fire in 1262/3 destroyed the map and most of the decoration, the walls had apparently been covered with depictions of the apostles, strange animals mentioned by classical writers and a painting of a genealogical tree showing the descent of Jesus from Jesse of Bethlehem by way of King David, known as the Tree of Jesse. There is also a reference to a 'history' which could well have been the map itself, as the encyclopaedic content of mappae mundi often led them to be described as histories. An inscription on the Hereford Mappa Mundi, for example, explicitly refers to it as an 'estoire'. Daniel Birkholz has suggested that the map at Westminster would have been placed behind the great bed, framed by the canopy as a sort of backdrop to what in effect was a throne.[71] A mappa mundi the size of the Ebstorf world map, however, could hardly have fitted into the available space at the top end of a bed. Even if it did, the difficulty in making out detail at a distance makes it more likely that the map was placed behind a lesser bed or couch positioned along a wall, of a type shown in a miniature of Christine de Pisan presenting her book to the queen of France in her bedchamber.

The decoration of the room served to emphasize the sacred nature of kingship and the monarch's role as God's anointed – and, implicitly, to bestow divine sanction on his secular policies and ambitions. Thus the divine became allied to the mundane in the setting of palaces as well as cathedrals. The details on the Psalter Map suggest that as well as the overall emphasis on God as benevolent master of creation, the map was designed to focus particularly on Henry's dominions, featuring a remarkably accurate depiction of England, and naming the Norman and Angevin dominions in France which Henry's father, King John, had lost. In that context, it thereby implied divine support for the justice of Henry's territorial claims.

The model for the enormous Ebstorf world map (p.80), destroyed during the Allied bombing of Hanover in 1943, may well have been similar in nature. Its content suggests that this model was commissioned by a duke of Brunswick. This could have been Otto the Child, Duke of Brunswick, in about 1238–9 for his palace in Lüneburg, or even Otto's uncle, Emperor Otto IV (who was also duke of Brusnwick) at the beginning of the thirteenth century. As Armin Wolf has pointed out, in addition to its encyclopaedic and deeply pious content the map exaggerates the size of the Duke's dominions in northern Germany and emphasizes the lands ruled by his relatives and allies at that time.[72] Alternatively, it has been argued that the imagery was equally applicable to the political situation in about the year 1288.[73] It seems likely that the map that perished in the Second World War was a slightly amended copy of the Lüneburg map, made by, or for, the nuns of Ebstorf in about 1300, just as the Psalter map was a copy of the earlier world map displayed in Westminster. The survival of copies of both maps and contemporary mentions

Christine de Pisan presenting a manuscript to Queen Isabelle of France, c.1420. British Library Harley MS 4431, fol.3

A chamber with a decorated cloth behind the couch, c.1470. British Library Royal MS 15. D.1. fol. 18

of Henry III's map suggest that they were reasonably well known in their time because they were on display – if only to highly select groups of visitors.

That the maps were reasonably well known is further supported by P.D.A. Harvey's recent discovery that, from as early as the time of Edward I, a world map painted on cloth was included in the travelling wardrobe of the kings of England – something that inventories of King Henry VIII's goods show was still the case in England as late as the 1540s. Whether in his bedchamber or elsewhere the king would have appeared in close vicinity to, and possibly in front of, a mappa mundi to emphasize his position as God's anointed, his personal piety and his own intellectual credentials. Smaller mappae mundi were to be found elsewhere in palaces.

The inventories of Henry VIII's goods of the 1540s mention that a mappa mundi was hung in the corridor leading from Henry VIII's private rooms in Hampton Court to the chapel. The other items on the walls were paintings on religious themes, suggesting that they were collectively intended to mark the transition from the secular to the spiritual and to prepare the king and his suite for their devotions.

The important role played by the bedchamber in the medieval palace of Westminster was a precursor of later developments in the architecture and layout of palaces. From the sixteenth century, and before that in Italy, great halls gradually lost their importance and eventually disappeared. As the state became ever more identified with the person of the monarch, so reception rooms tended to become focused on the ruler's person. The more private the space, the greater the honour for whoever was allowed access. The most public of these 'private' chambers was the so-called 'presence chamber'. This might also be used as a state dining room where the king, seated under the canopy of state, ate in public and was served by particularly favoured courtiers, or it could be used for semi-public receptions and state ceremonies.

Such rooms were sometimes the location for map murals, as was the case with the *Sala Bologna* in the Belvedere villa in the Vatican. In the course of the 1520s Francis I of France (presumably inspired by the galleries of town views of late fifteenth-century Italian palaces and ultimately by the *Forma Urbis Romae*) is known to have commissioned a vast bird's-eye map of Paris, estimated to have measured about 5 metres by 4 metres, for the walls of one of his palaces, perhaps the Louvre in Paris.[74] It is possible that the map may have graced the privy chamber – which was more private than the presence chamber and was meant for the reception of particularly important visitors. It was for this room that Holbein created for Henry VIII the great fresco showing the Tudor dynasty, for which a cartoon survives; the fresco itself was recorded in a painting by Remigius Leemput in the 1680s before its destruction by fire in 1698. It is possible that the walls of that room also accommodated a large view of London by Anthonis van den Wijngaerde for which preparatory drawings – now in the Ashmolean Museum in Oxford – still survive.[75]

It was not only town maps and views that adorned reception rooms, however. Before its destruction by fire in 1793, the walls of the dining room at Cowdray Park in Sussex, which seems to have been decorated in about 1550, contained mounted pictorial maps of Henry VIII's Boulogne campaign of 1544–5. These were probably copies of a larger series of map-like history paintings – which also probably showed the Field of the Cloth of Gold and other Henrican 'triumphs' – that had been painted for the private apartments of the Palace of Whitehall in the 1540s.[76]

Ironically the one room in the private apartments of a palace not adorned with maps by the late seventeenth century was the state bedroom, although it continued to play the central part in court ceremonial that it had in fourteenth-century England.

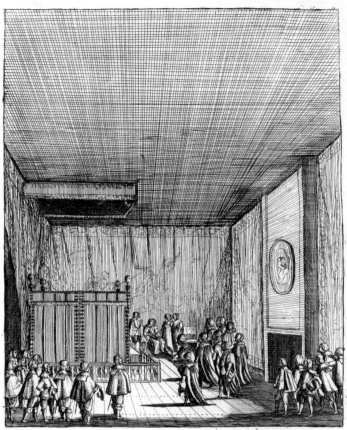

COMME LE MY LORD MAIOR ACOMPAIGNEDE SES COLLEGVES VIENT SALVER LA REYNE LVY FAIRE SES PRESENS

The bedchamber/reception room of Queen Marie de Medici in St James's Palace, 1638. From *Histoire de l'Entree de la Reyne Mere du Roy Tres-Chrestien dans la Grande Bretaigne, etc,* London, 1639, British Library, c.37.L.9

70 For what follows see Paul Binski, *The Painted Chamber at Westminster* (London, Society of Antiquaries, 1986) and Birkholz, pp.5–11.

71 Birkholz, pp.3–19.

72 Armin Wolf, 'News on the Ebstorf World Map: date, origin, authority' in Monique Pelletier (ed.), *Géographie du Monde au Moyen Age et à la Renaissance* (Paris, Comité des Travaux Historiques et Scientifiques, 1989), pp.51–68.

73 Jürgen Wilke, *Die Ebstorf Weltkarte*, 2 vols. (Bielefeld, Verlag für Regionalgeschichte, 2001).

74 Jean Dérens, 'Les Plans Généraux de Paris au xvie siècle', in Michel Le Möel (ed.), *Paris a vol d'oiseau* (Paris, Délégation à l'action artistique de la ville de Paris, 2000), pp.24–47. This was the model for the map tapestry commissioned by Cardinal de Bourbon and the gouache plan prepared for the magistrates of Paris.

75 See Barber in *Inventory* iii (forthcoming); for Wijngaerde's drawing see Howard Colvin and Susan Foister, *The Panorama of London circa 1544 by Anthonis van den Wyngaerde* (London, London Topographical Society, 1996).

76 See Barber in *Inventory* iii (forthcoming).

A royal wall map in miniature?

The 'Psalter' world map takes its name from the book of psalms, or psalter, in which it is to be found. Despite its diminutive size, the Psalter world map has long been recognized as one of the 'great' medieval mappae mundi, and it gives a reliable impression of their original colourful appearance. Much of the detail, such as the line of the Nile running atop the images of the marvellous creatures on the fringes of the Earth, is barely visible to the naked eye, strongly suggesting that the map is a reduction from a much larger original.

The overall structure of the Psalter map itself is similar to both the Ebstorf and the Duchy of Cornwall maps. All are distorted to give space to features that the creators judged to be important and their geographical outlines seem incomprehensible to most modern viewers. The strangeness is increased because they have East, not North, at the top (though the word 'orientation' stems back to the times when this was normal). Each map was slightly different. Unlike the Ebstorf map, where God becomes the world, and the Hereford map, with its emphasis on the Last Judgement, God in the Psalter map holds a T-O-shaped orb in his left hand and blesses the world, flanked by two incense-swinging angels. On the back of this map, however, is another more schematic one. The surrounding imagery in this map shows God holding the world in a manner that recalls his figure on the Ebstorf map.

The content of the Psalter map reflects the usual elements in encyclopaedic medieval world maps. There are plentiful references to the Roman Empire, the Alexander romance, the empires and ages of the world and to events and persons recorded in the Bible. The right edge of the map is filled with monsters and marvels, derived ultimately from Herodotus and Pliny via Isidore of Seville. The monsters are in almost exactly the same sequence as those on the Duchy of Cornwall and the Ebstorf maps. The indication of the important contemporary commercial cities of Paris, Lyon, Barcelona, Cologne, London and perhaps Salzburg – along with references to the Crusades, particularly Damietta captured by the crusaders in 1219 and 1249 – reflect preoccupations of the thirteenth-century world.

The focus of the original of the Psalter map seems, however, to have been England. The miniaturized outlines are amazingly realistic, with a recognizable Scotland, Wales and Cornwall. They are similar to those on much earlier English maps, and may ultimately derive from a Roman source. The Thames and Severn are shown, with London indicated by a golden dot. The naming in such a small map of Normandy and Aquitaine, lost ancestral French dominions of Henry III of England, is particularly significant. The source map probably contained the names of other Plantagenet dominions in France such as Anjou – names which would have had a particular relevance for the English king. It is hard not to see the Psalter map as a close copy of the lost great world map which adorned Henry III's bedchamber in Westminster Palace from the mid-1230s; it is certainly known to have been much-copied. The fact that the book and the maps within it were created in a London workshop strengthens this hypothesis.

Barber and Brown (1992), particularly pp.31–33; Morgan (1988), ii, no.114, pp.82–5; Barber in Harvey (2006), pp.15, 17–19.

'Psalter' world map, Westminster, c.1265
Manuscript on vellum, 9 cm diameter (map)
British Library Add. MS 28681 fol. 9

A fragment of a royal mappa mundi

This fragment of a once-magnificent mappa mundi shows only the south-western part of Africa, or less than a quarter of the original. It has been estimated that the diameter of the complete map would have been 164 cm, making it slightly larger than the Hereford Mappa Mundi. It was almost certainly drawn in the last quarter of the thirteenth century by an artist attached to the court of Edward I of England. It was presented to the College of Bonhommes in Ashridge, Hertfordshire, an Augustinian religious house, shortly after 1283 – possibly by the College's founder, Edward I's cousin Edmund, Earl of Cornwall (d.1300). It was probably accompanied originally by a copy of Peter Comestor's *Historia Scholastica*, now in the British Library (Royal MS 13.D.VI), which has illustrations by the same hand. The magnificent map would have served to emphasize the new college's prestigious royal connections, and may even have served as a backdrop for the public audiences of Edmund, who frequently visited the college and indeed died there.

Like the Hereford Mappa Mundi, the Duchy of Cornwall map highlights the contrast between human and divine time. It contains five (out of a possible original total of nine) roundels along its lower edge illustrating the ages of man (and woman), the usual seven being supplemented by one showing death/purgatory and another showing resurrection (as an angel). However, the Duchy and Hereford maps belong to quite distinct design traditions.

There is considerable evidence to suggest that the ultimate model for the Duchy of Cornwall map, like the Psalter map, was the great and much-copied world map of Henry III at Westminster, destroyed by fire in 1263. The map at Westminster may also have shared a common source with the Ebstorf map. All followed the guidelines for map construction set down in the twelfth century by Hugh of St Victor, and the structure of southwest Africa on the Duchy and Psalter maps, for example, is almost identical as are many other features: the sequence of 'Wonders of the East' on the Duchy, Psalter and the late fourteenth-century Aslake maps is the same, except in small details, running (from the bottom or west beyond the Hyperian mountains) from the ferocious dog-headed cannibals (the Cynomolgi) and the cannibal Anthropophagi to the Himantipodes (wrongly named Archaliotitte on the Duchy map) who creep on all fours, the Troglodites, cave-dwellers who run so fast that they catch beasts by leaping on them, and an unnamed race, like the Blemyae, whose heads grow beneath their shoulders. The association of both the Duchy and the Ebstorf maps with the works of Peter Comestor, who theorized about the nature of the world as the body of Christ, suggests that like the Ebstorf map and one of the Psalter maps, the Duchy map may originally have shown the world as being held by or forming part of the body of Christ.

Ashridge College was dissolved in 1539 and the map came into the hands of the Court of Augmentations. It was probably destroyed and its parts re-used as bindings for administrative records a few years later, during the iconoclastic reign of Edward VI.

Harvey (2006), pp.19–23; Haslam (1989), pp.33–44.

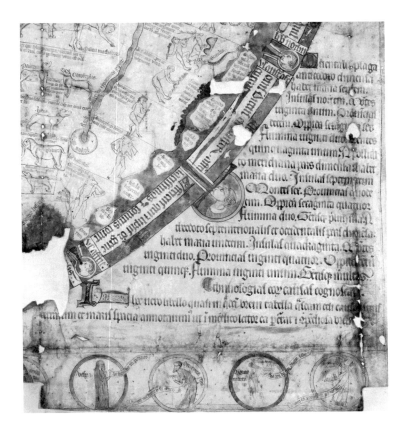

Duchy of Cornwall map fragment, Westminster (?), *c*.1290
Manuscript on vellum, 62 x 53 cm
Duchy of Cornwall Office, London

Brunswick and the world

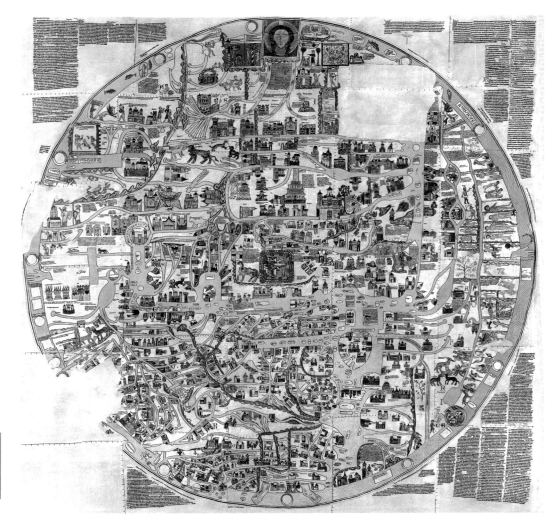

Anon, *The Ebstorf World Map*, *c*.1300,
Ebstorf, Kloster Ebstorf

Photographic reconstruction
(originally manuscript on vellum),
357 cm in diameter

The Ebstorf world map, the largest known medieval mappa mundi, was discovered in Lüneburg in Germany in 1830 and destroyed by Allied bombs in 1943. Luckily it was photographed in detail in 1891 so its appearance – apart from the square that seems to have been cut out shortly after its discovery and which has not yet been found – can be accurately reconstructed.

The Ebstorf map's content owes much to the same encyclopaedic writings of classical Roman and early Christian authors, and to the Bible, as all the other great maps. Like all medieval mappae mundi, it also includes more modern information such as some important contemporary commercial and pilgrimage routes. Its structure and the way in which its information is organized resembles the Psalter and Duchy of Cornwall rather than the Hereford world maps and, like the Duchy of Cornwall world map, it may have been particularly influenced by Peter Comestor's *Historia Scholastica*.

It may be significant that both the Duchy of Cornwall and the Psalter maps seem to have had royal associations, since the prototype for the Ebstorf world map probably did too. A large part of Europe, in the bottom-left quarter of the map, is taken up by northern Germany, and the symbol for Lüneburg, the principal seat of the powerful dukes of Brunswick, is surpassed in size only by those for Jerusalem, Rome and the Tower of Babel. Recent writers have also noted that particular attention is paid to the possessions of the Duke of Brunswick and his relatives and allies, though there is some dispute as to whether this applies to the situation in about 1210, 1238–9 or after 1288.

The writers agree, however, that the prototype could have been displayed in one of the palaces of the dukes of Brunswick and that it is likely to have been copied in about 1300 for, and quite possibly by, nuns of the convent of Ebstorf ('Ebbeskesstorp', bottom left). The copyists probably added a few features, such as the mention and depiction of the local martyrs' graves, but on the whole they seem to have reproduced the map before them. Given the frequency with which world maps were copied in the Middle Ages, it is at least possible that one of the sources for the Ebstorf world map's prototype may have been Henry III's map at Westminster or the model for it. A possible channel may have been Edward I's brother Richard (father of Edmund, Earl of Cornwall), who, between 1256 and 1272, was elected king of the Romans, and as such heir to the German emperor. If the prototype for the Ebstorf map is much earlier, the link could have been the English cleric Gervase of Tilbury who worked for the dukes of Brunswick in his later years.

The spiritual message of the Ebstorf world map, like the Psalter map, is more benevolent than the austere message of the Hereford Mappa Mundi. In accordance with the writings of the influential twelfth-century theorist Hugh of St Victor, the world is shown as the body of Christ, whose arms and feet emerge from behind the map. In contrast to the Hereford map's emphasis on Christ as Judge, this implies Christ's involvement with, and understanding of, humanity and the world. In the context of Lüneburg it also would have expressed the hope that the Duke, his relatives and friends enjoyed a special place in God's affections, and in the secular as well as spiritual scheme of things.

Kugler, Glauch and Willing (2007).

New wine in an old bottle

Grayson Perry has written of this map: 'Most of my art stems from seeing an artefact in a book or a museum and then making my own version of it. I make things that recognizably belong in traditional categories, pots, costumes, tapestries and maps.

'The formal layout of *Map of Nowhere* came from the Ebstorf map, a 13th-century mappa mundi by Gervase of Tilbury destroyed by Allied bombing in 1943. I liked the idea of Christ's body somehow encompassing the limits of the known world, it also echoes Leonardo da Vinci's famous drawing of Vitruvian man. The title came from Thomas More's *Utopia* which was a pun on the Greek *ou topos*, meaning no place.

'I wanted to make a map of the beliefs, head-lines, clichés and monsters that populate my social landscape. At the top of the map, to the right hand side of my head, is a swirling whirlpool labelled *Baloney Generator*. This was cognitive scientist Stephen Pinker's phrase, describing our brain's desperation to make sense of the world and its often spurious rational-izations of our intuitive behaviour. Perhaps this map is an attempt to chart a meandering journey through my own psyche and contemporary life. In the centre of the map sits a figure in a cage behind a wall being visited by a stream of pilgrims on a small island labelled *Doubt*. I am distrustful of those who are vehemently sure about their beliefs. Above the island of *Doubt* in the sea of *Despair* is a ship labelled *The Sadness of the Excessively Logical*; below it on the shore is a phallic rock dwelling, *Know All*.

'Making this print took several weeks spread over six months. First I roughed out the overall compo-sition showing my limbs and torso as rivers and lakes, then I started in the top left-hand corner and worked my way across, deciding on the detail spontaneously as I drew. This piece reflects my concerns at the time, late 2007–early 2008, about everything from class and turbo-consumerism to green politics and intellectual snobbery. The names of the various sites are often current media buzz-words and phrases: *Kidults, Neighbours from hell, Shopaholic, Thumb genera-tion, Faith-based intelligence, the Bilbao effect.*

'Traditional features of an old map, beasts and the weather, are given a sociological equivalent. A cat-shaped lake, *Nature*, in the lower left quarter is surrounded by a forest within which are hunting lodges called such things as *Middle class guilt, Tree huggers* and *Carbon footprint*.

'In the very top right-hand corner, instead of a cherub blowing the east wind, a female demon radiates the toxic unconscious messages of the working-class parent, *Not for the likes of us; Got above ourselves, have we?*

'Below the flat disc of this anatomical cosmos I have drawn in strong perspective a Ruritanian fantasy of pilgrimage. Here the exotically dressed faithful congregate at the last staging post before heading off up to the mountain-top shrine. Their goal, a monastic building, is lit against the stormy sky by a ray of heavenly light originating from my anus.'

Grayson Perry, *Map of Nowhere*, London, Paragon Press, 2008
Etching on paper, 153 x 113 cm
British Library Maps CC.6. a. 26

THE CABINET

As early as the 1420s Niccolò Niccoli – a leading Florentine humanist associated with the translation into Latin of Ptolemy's *Geographia*, who took a particular interest in its lists of place-names as a means of understanding ancient history – housed his maps of Italy and Spain and possibly a great world map, together with his greatest treasures, in a special room.[77] His example was followed by such influential figures as the banker and statesman Cosimo de Medici 'the Elder', the effective ruler of Florence in the middle years of the fifteenth century. From them this classification of fine maps as prestige objects to be accommodated with other valued items spread throughout Italy, and eventually beyond.[78]

As with the other rooms in a palace, the space where a ruler stored and displayed his greatest treasures was known by a variety of names over the centuries. If small it was called a closet, and the inventories of Henry VIII's goods mention that he had a small 'mappa mundi' in a 'closet' in Greenwich Palace. As late as 1679 the English mapseller John Garret was advertising some of his maps as being suitable for 'closets'.[79] In Italian homes these smaller spaces were sometimes called '*studioli*' – a term taken over in England by the 1540s. Here the space was called a study, and was also associated with work at a desk.

If the room was large and could serve as much for storage as a display area, it tended to be called a wardrobe, also known by its Italian name of *Guardaroba*. The most famous cartographic wardrobe was the *Guardaroba Nuova*, also known as (yet another) *Sala del Mappamondo* in the Palazzo Vecchio in Florence. This was constructed for Cosimo I de Medici, Grand Duke of Tuscany, with assistance from the artist and theorist Giorgio Vasari; the maps were thought up, edited and painted from 1563 by Egnazio Danti and, after his dismissal following Cosimo's death, by Stefano Bonsignori.[80] A variety of 'wardrobes' and their contents are listed in the inventories of Henry VIII, and they were also to be found in the palaces of his fellow-rulers. Eventually these evolved into treasuries – or, in German, *Schatzkammer* ('*Schatz*' being the German word for treasure) or *Kunstkammer*, which makes the room's links with sophisticated art ('*Kunst*' in German) explicit.

While a private house would have had only one room in which to display the owner's treasures, in most palaces there were several – both the king and his queen had at least one in their separate wings. The Vatican, not surprisingly in view of the papacy's enormous wealth that attracted criticism from Martin Luther and other less celebrated critics, had a whole building that effectively served as a repository for art and treasure. This was the Belvedere villa, built in the grounds of the Vatican in the late fifteenth century as a summer house where popes could relax in the midst of their gardens and indulge in connoisseurship. It grew in size over the following 150 years and came to include all the cartographic rooms and galleries in the Vatican: the loggia with town views by Pinturrichio, the *Sala Bologna*, the *Terza Loggia* and the *Galleria delle Carte Geografiche*.

Whatever their number, name or size, these rooms were situated in the most private areas of the ruler's private apartments; access to them was thus regarded as a particular privilege and a sign of special favour. By 1700, reflecting France's cultural dominance, such rooms had commonly come to be known across Europe by the French word 'cabinet' – from the piece of furniture in which the royal treasures were stored and displayed. To this day the king of Sweden receives ambassadors for their first official audience in a small cabinet at the far, and most inaccessible, part of what were once the private quarters of Swedish rulers in the royal palace in Stockholm. Because of its connotations as the room where the monarch met his, or her, most favoured advisers in conditions of total confidentiality, 'cabinet' took on its modern political meaning as the most important committee of government.

Despite the connotations of political power attached to the room's location, the contents of such rooms had no ostensible connection with either

The library, Ham House, London with cartographic fire screens of 1743

The *Guardaroba Nuova*, Palazzo Vecchio, Florence

politics or power. They were intended instead to present the ruler as a cultured, well educated, well informed and intellectually interested individual whose personal qualities suited him (and very rarely her) to be a head of state. This was particularly important for rulers of minor states. In their cases, the display of the most up-to-date world maps reflected not claims to territorial domination, but rather a reputation for universal knowledge. Over and above this, but buttressing the benevolent image, and to to show something of his private side, the ruler might also display objects of personal importance, such as religious relics or souvenirs of people they had known or places they had visited in their youth. Monarchs might also want to demonstrate their historical knowledge or piety through ownership of particularly accomplished, as well as aesthetically pleasing, maps depicting the ancient world or the Holy Land. Inventories, memoirs, contemporary correspondence and a few prints and paintings show that cabinets also often contained large and miniature maps, globes and gold, silver or gilt armillary spheres. They might also include surveying instruments, clockwork globes and globes that could be used as drinking vessels or as clocks, along with the intricately crafted gold and silver vessels, the jewellery and the natural history specimens, curiosities and oddities from outside Europe.

As already mentioned, the primary reason why maps and related objects were to be found in cabinets was because prior to about 1780 maps were valued for the encyclopaedic information they contained. They were not seen simply as geographic aids. They could be used, as was the case with those on the front of cupboards in the *Guardaroba Nuova*, as a visual index to the places of origin of the treasures they held within.[81] Maps might also show the places where famous historical (and also recent) events had taken place (often with pictures of the relevant scenes) and the locations of historic cities, such as Babylon, Jerusalem, Rome and Alexandria. They would contain miniaturized depictions of modern cities and the best-known sights, of the world's flora and fauna, of the costumes worn by the different social classes and the clothing and customs of distant regions.

The maps often also contain information and representations of the political constitutions of the lands they portayed. To fulfil this role effectively, they had to be up-to-date.[82] Mapmakers usually had to collate information from several sources, including other maps, travel accounts, book illustrations and occasionally confidential reports – if only to fill the amount of space available on large, multi-sheet wall maps, which in these respects presented far greater challenges to the mapmaker than the smaller maps designed for binding into atlases. The maps to be found in cabinets were therefore sometimes the most advanced currently available.[83]

It was only in the last quarter of the eighteenth century, at the height of the Enlightenment and as geography emerged as a separate academic displine, that perceptions radically changed. Henceforth maps came to be esteemed primarily for their geographical accuracy and scientific precision – to the exclusion of other elements, which were dismissed as frivolous decoration detracting from the maps' essential (geographical) purpose.[84] In the same decades, royal collections of curiosities in old-fashioned cabinets were dispersed, their contents in many instances being transferred to the first modern national museums, such as the Louvre in Paris or the Uffizi in Florence.

77 Fiorani (2007), p.806, no.12; Dalché (2007), pp.293–4.
78 David Woodward, *Maps as Prints in the Italian Renaissance: Makers, Distributors and Consumers* (London: British Library, 1996), pp.88–93.
79 Reproduced in R.A. Skelton, *County Atlases of the British Isles 1579–1850. A Bibliography* (Folkestone, Dawson, 1970), pl.27.b.
80 For a recent and full discussion of all aspects of the *Guardaroba Nuova*, see Fiorani, *Marvel of Maps* pp.21–136.
81 Fiorani, 2005, pp.87–9.
82 Fiorani, 2005, p.188ff.
83 Fiorani, 2005, pp.105–32.
84 These new, plain 'geographic' maps, however, met the austere aesthetic canons of the then fashionable neo-classicism, and in overall appearance some bear comparison with the best graphic art of the period.

The Zurich municipal cabinet of curiosities in the Wasserkirche contained numerous maps and globes

The world for a king

This is the grandest of three large world maps created by Pierre Desceliers (1487 (?)–aft.1552). All were dedicated to French leaders, but this particular example was presented to the king himself, Henri II, whose arms are shown at the bottom left and whose crowned initials adorn the left and right borders. The other arms are those of Anne de Montmorency and Claude d'Annebault, constable of France and admiral of France respectively. Montmorency was Henri's first minister and his arms are surrounded by the chain of St Michel, the premier order of French chivalry. By contrast d'Annebault was out of favour, and his arms are shown unadorned. There is an empty space for another coat of arms at the top left, perhaps intended for a courtier who had just fallen from favour.

Desceliers was the most distinguished French chartmaker of the Portuguese-influenced Dieppe school of hydrography, and has been considered the father of French chartmaking. A priest, he also worked as chief pilot to the French navy at Dieppe and was responsible for certifying the reliability of the charts made by other chartmakers. In addition, Desceliers produced atlases and books of instruction on geography, so no better person could have been selected to prepare this prestigious chart. It is likely that an unrecorded professional artist drew the figures, ships and animals.

The chart is a compilation of ancient and modern information. In Canada there is a depiction of a battle between cranes and pigmies, the story of which goes back to Homer. This scene is normally placed in Tartaria, and its presence in North America suggests that Desceliers still thought that America could form part of Asia. On the other hand there is information about the recent voyages to Canada of Jacques Cartier and the Sieur de Roberval (1534–42) and the Canadian coastlines reflect their discoveries. In South America the Spaniards are shown attacking the Inca empire in Peru. The twenty-five text panels contain ethnographic information, and details about the locations of precious spices and minerals derived from direct observations by Norman and Portuguese sailors. Despite the apparent confidence of the coastal outlines and the ethnic depictions, caution is the keynote. Desceliers skilfully omits most of the Pacific, last traversed twenty-five years earlier and still largely unknown. The ethnic depictions in the great southern continent mainly show peoples

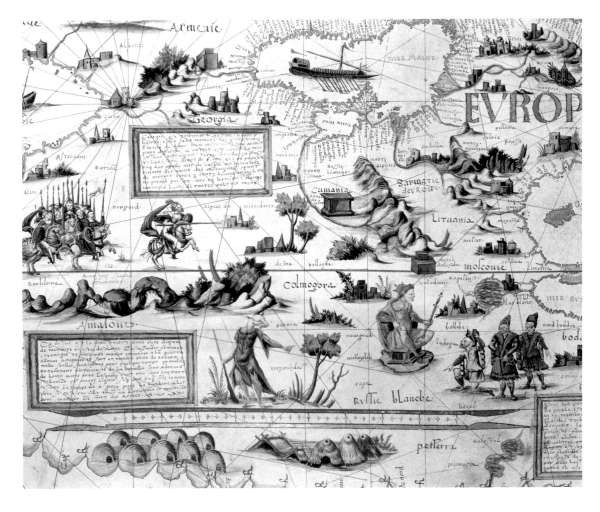

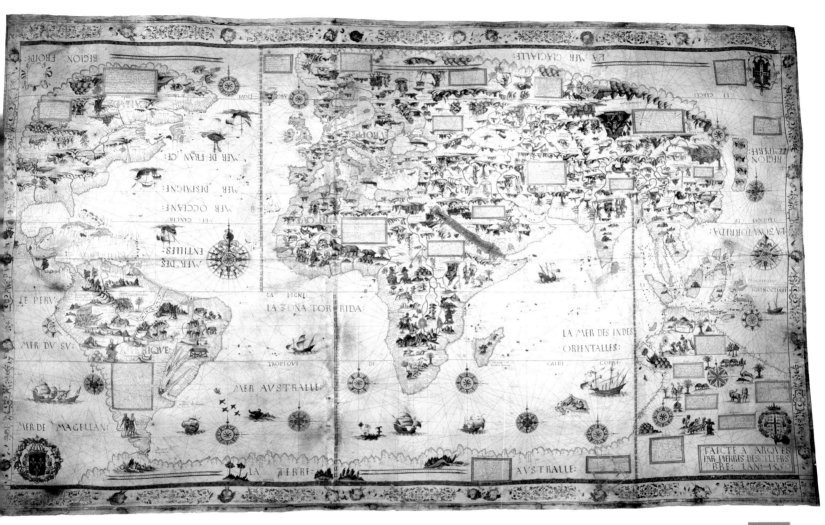

Pierre Desceliers, [*World map*], Arques, Normandy, 1550
Manuscript on vellum, 139 x 219 cm
British Library Add. MS 24065

and scenes from elsewhere; they seem to suggest merely that the continent, admitted to be unknown, might be inhabited.

The map was too cumbersome and on too small a scale to have been used for strategic, administrative or military purposes. As a luxurious and expensive manifestation of knowledge, however, it was suitable for display in a cabinet. Indeed, a later Dieppe chartmaker, Guillaume Levasseur, frankly confessed in 1608 that 'all the [Norman] universal maps serve more for decoration, like paintings, than for instruction and information'. This map has no orientation and the way the place names are written indicates that it was was meant to be viewed from north and south, presumably on a table. King Henri II, who is known to have taken a great interest in exploration, would undoubtedly have been delighted by it.

The map's later history is largely unknown. It was first described in 1852, when it was owned by Professor Cristoforo Negri, an academic and politician later to be the founding president of the Italian Geographical Society. In August 1860 he offered it to the British Museum (for $250). The museum, under pressure from the Chief Librarian Sir Antonio Panizzi and R.H. Major, an expert in the history of maps and travel, purchased it in March 1861 for $150.

de Challaye (1852), pp.235–44; du Jourdain, de la Roncière, Azard, Raynaud-Nguyen and Vannereau (1984), pl.47, pp.231–2; Toulouse (1998), pp.52–5; Toulouse (2007) iii, pp.1550–67; acquisitions archives of Department of Manuscripts British Library; Diary of Sir Frederic Madden, (4, 24 October, 31 December 1860, 12 January 1861).

A renaissance mash-up

There is probably no map which more perfectly embodies the motivation behind Renaissance principles than the 1551 map of Rome by Leonardo Bufalini, or which transmits the power most appropriate for the inner sanctum of a ruler's world: intellectual power. Bufalini's map, the first large-scale map of Rome to have been made since antiquity, was well placed to record a city changing under the urban schemes of successive popes wishing to stamp their names on its infrastructure. At the same time by overlaying the modern city over a plan of ancient Rome at the height of its power, Bufalini's map enabled the viewer to compare the antique and the contemporary in a fresh and novel way, and to reflect upon humanistic concerns prevalent among the intellectual and ruling European elite.

It is difficult to overstate the influence of ancient knowledge in the Renaissance, or the attraction it held for figures in positions of authority. Antiquarian learning, studying the ancients in order to bridge the intervening dark ages, was a pioneering undertaking similar to modern science. And, like modern science, continuous breakthroughs were made in the form of the discovery of architectural remains, sculpture and pottery (and fragments of the huge *Forma Urbis Romae*, of which the similarities with Bufalini's map may, or may not, be purely coincidental). In addition, classical texts and treatises were by the sixteenth century being used and updated by artists and architects. These sources, and the understanding of them, served to validate and to order the Renaissance world, offering a blueprint of perfection for those with the means to recapture it. By collecting such objects as maps, texts and fragments of antiquity, symbols of learning and prestige, the owner could align himself ultimately with ancient Rome.

Bufalini was a military surveyor who had been employed on strengthening Rome's fortifications after its sack by Habsburg troops in 1527 (an event that was met with Europe-wide condemnation). He was well placed to produce a contemporary plan using the skills of his trade. Bufalini's knowledge of ancient Rome, however, appears to have been gained partly from the remains which still survive today, to which he added information derived from texts, with a fair amount of embellishment. Throughout the sixteenth and seventeenth centuries bird's-eye views of ancient and modern Rome were produced on separate sheets, but Bufalini's idea of producing the map as a plan, rather than the more fashionable 'bird's-eye' view, made it possible to overlay one vision of the city over another. Despite the fact that only two examples are known today, Bufalini's plan does not appear to have been a sufficiently elegant image to satisfy the tastes of the elite. As an intellectual document, however, it remains a tour-de-force.

Maier (2007), pp.1–23.

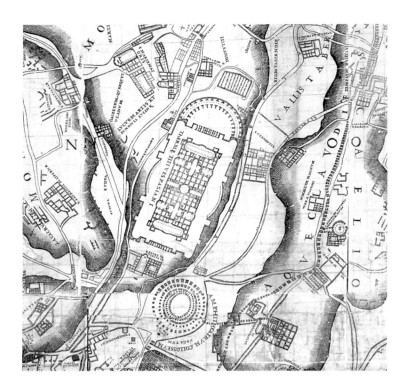

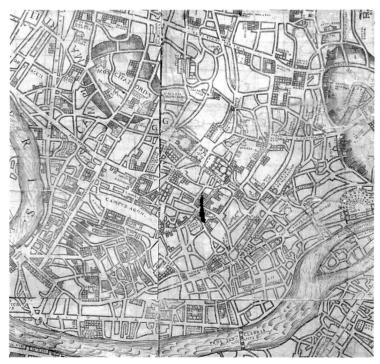

Leonardo Bufalini, *Roma*, Rome, 1551 (1560)
Woodcut on twenty-four sheets, 206 x 189 cm
British Library Maps S.T.R.

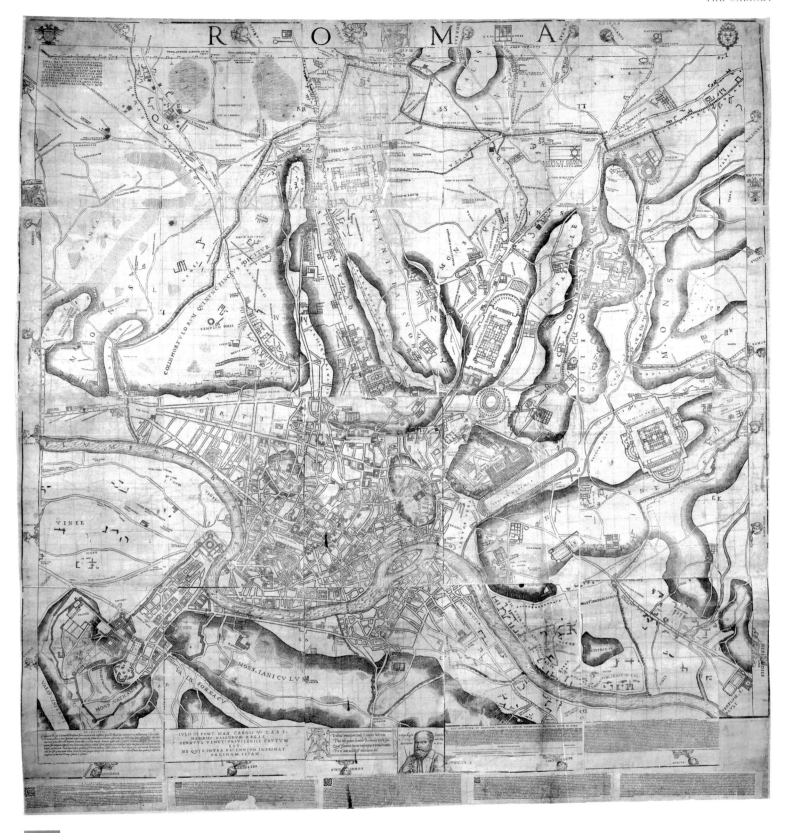

A renaissance underpinning

With its spacious design and varied content, this first map by Abraham Ortelius (1527–98), dated 1564, is markedly different to many of the densely packed, encyclopaedic world maps produced at the same time. Yet its blend of intellectual comment and esoteric elements would have spoken clearly to a learned audience familiar with ancient precedents and Renaissance discoveries. The map is not dedicated to any one of the European rulers who would doubtless have been keen to associate themselves with such a popular author and compiler as Ortelius. Instead, the dedication is to the author's fellow antiquary and humanist Martinus Luvinius, and it is precisely this apolitical stance that enabled the map to transcend national power, and in doing so reinforce the intellectual reputation of the European ruling elite.

The idea that classical knowledge – the advances in science and art made by the Greeks and Romans – formed the basis of Renaissance

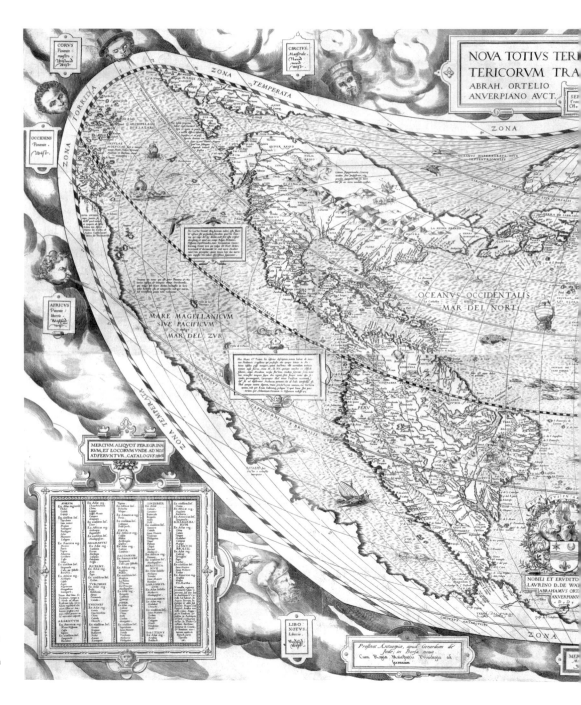

Abraham Ortelius / Gerard de Jode, *Nova Totius Terrarum Orbis Juxta Neotericorum Traditiones*, Antwerp, 1564
Copperplate engraving on eight sheets, 109 x 180 cm
British Library Maps C.2.a.6

knowledge is represented in the map through a series of comparisons between the old and the new. The Greek mathematician Claudius Ptolemy had struggled with the concept of portraying the three-dimensional world onto a flat surface, and the striking projection used by Ortelius after Bonne contributed another possible solution in the years prior to Gerard Mercator's famous projection of 1569. The ancients, of course, had no knowledge of the New World discovered by the Portuguese and Spanish in the fifteenth and sixteenth centuries, and would not have witnessed the extraordinary form of the cities of Cuzco and Mexico City, here afforded special prominence on the map. Ortelius's statement that 'the ancients had no knowledge of Africa' does not suggest that the ancients were incorrect (in fact, Ortelius uses Greek geography for the interior of Southern Africa) – simply that they lacked the complete picture now available to the intellectual society of early modern Europe.

Ortelius's map would have fulfilled a further function in the study or private chambers of a ruler. It would have constituted a geographical index for the collected objects, such as cinnamon, ruby and rhubarb, which would also have been kept there. A listing is provided of some of the more unusual and valuable objects, along with indications of where they may be found. Rhubarb, for example, is believed to have been brought to Europe by Marco Polo in the thirteenth century. By owning these things, and showing knowledge of their origins, a ruler could position himself in the vanguard of understanding and discovery – reinforcing his place within it as he did so.

Ortelius's map was published in Antwerp by Gerard de Jode (1509–91), and two surviving copies are known. After compiling the first atlas in 1570, Ortelius produced the first atlas of classical geography, the *Parergon*, in 1579.

Schilder (1987), pp.3–58; Shirley (2001), p.569 (no.575).

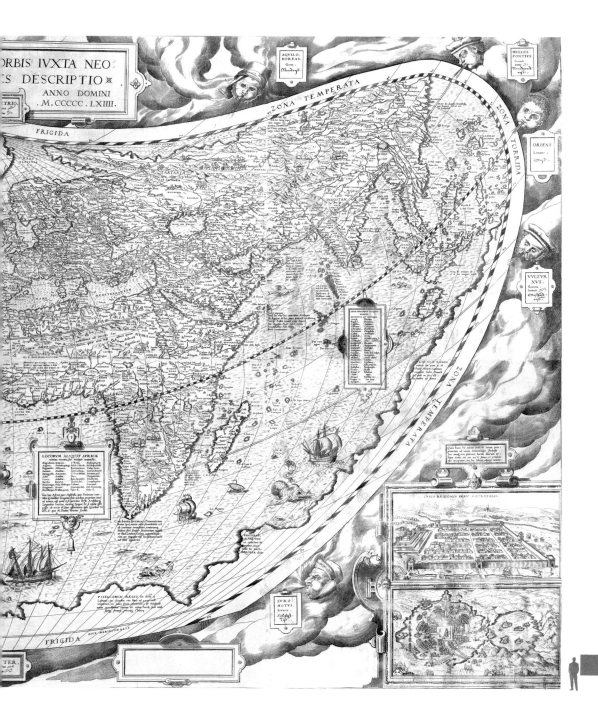

A souvenir

John Evelyn (1620–1706), the famous diarist and print collector, was one of the earliest Englishmen to embark upon a Grand Tour. From the seventeenth to early nineteenth centuries the Grand Tour was to become a staple 'rite of passage' for young men of noble birth or heritage; it offered a chance to travel through Europe and its major cities, to learn to experience works of art in accordance with contemporary tastes for antiquity and knowledge, and, of course, to collect. Many English private collections today have as their foundations art, sculptures and objects of curiosity collected during travels in Europe.

Mannerist paintings, as well as sculpture and prints bought or plundered from France and Italy, came to be housed in long galleries or dining rooms of stately homes, but the cabinet or study offered a more private place where items of an altogether more personal nature might be kept. This map of Paris was bought by Evelyn in November 1643, as his annotation in a blank cartouche attests. He had arrived in Paris at the start of his tour that same month, aged twenty-three. The map itself is an interesting derivative of a famous early sixteenth-century bird's-eye view of Paris from the northeast. It was published by Anthonie de Vuauconsains in 1616 and includes an equestrian portrait of Henri IV of France (1589–1610), assassinated six years previously. This map honours a king who had converted from Protestantism to Catholicism in order to gain control of the city. Henri IV's scheme to beautify Paris included the creation of the *Place du Roi* (now Place des Vosges), the earliest Italian-style piazza north of the Alps, which was eventually completed two years after his death.

Thirty years later, when Evelyn acquired this copy, the map and its commemorative function would have been decidedly out of date. In fact the map was updated that same year by Nicolas Berey, who removed the equestrian portrait. In his diary of December 1643, perhaps while looking at this map, Evelyn notes Paris's large and round shape. Evelyn, in fact, appears more taken with the rather idealized representation of the city than its reality, likening a particular Parisian odour to sulphur 'mingled with the mudd.'

Boutier (2002), pp.122–3.

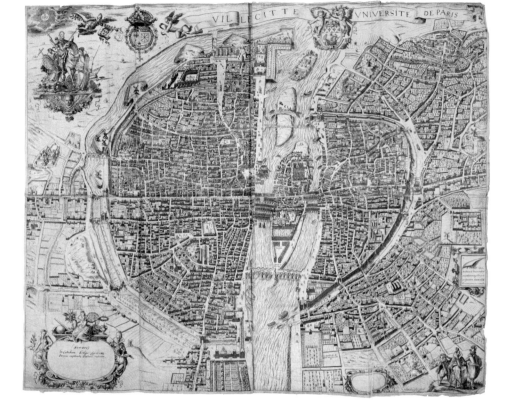

Jan Ziarnko, *Ville Citte Universite de Paris*, Paris, 1616
Copperplate engraving, 67 x 78 cm
British Library Maps CC.5.a.500

A reminder of home

George I was deeply attached to Hanover, the capital of his German lands, and he could well have displayed this map in his private quarters while he was in England.

The town of Hanover was hemmed in by its fortifications, but the land in the vicinity offered space for grandeur. This was particularly important for the great German court that Hanover had become since 1692, when George I's father had been elevated to the top rung of German princes as an elector of the Holy Roman Empire.

The electors of Hanover never managed to create the grandiose palaces aping Versailles that were being built elsewhere in Germany around 1700. However, George I's mother, the redoubtable Electress Sophia, was inspired by the Dutch gardens that she remembered from her childhood, and had expanded and embellished the gardens at nearby Herrenhausen, first laid out in 1666. Sophia's gardens became one of the horticultural and cultural wonders of Europe.

Sophia was assisted by the great philosopher Gottfried Wilhelm Leibniz, responsible for thinking out the complex allegories in which the gardens abound, and it was in these gardens that the music of the Hanoverian court composer, Handel, was first performed. Among the gardens' most impressive features were an outdoor theatre (C) and a jet of water that shot up to a hitherto unparalleled height of 35 metres (D). It was in the gardens that Sophia died in August 1714, just weeks before she would have become queen of England. The palace itself (I) was to remain a modest affair, and was destroyed by bombing during the Second World War. The old Orangerie (K), built in 1694–8, was, and remains, far grander, and it was here, transformed into a gallery, that Sophia lived. A new orangery (L) was built in 1720.

This plan can be dated to 1726–7, and probably commemorates the planting of the large avenue of lime trees, commissioned by the king to link Hanover to Herrenhausen in 1726. The absence of a numeral in the royal monogram above the title suggests that the plan was created before George I's death in June 1727. Herrenhausen can be seen at the top, with letters identifying the most important features of its gardens. Immediately above is the little palace that George had just presented to his illegitimate daughter, the Countess of Delitz. Between Herrenhausen and Hanover are another royal palace, Mon Brillant (U), the formal gardens of leading courtiers (R–T) and, closer to the town, the gardens of its burghers.

The plan of Hanover reveals what a tolerant and cultured place it was. Although it was officially a Protestant town, the map shows a Catholic church and a Jewish street, with a Jewish cemetery just beyond the walls. The archives are also identified on the map, as is a laboratory. Eighteenth-century British politicians tended to be rude about Hanover when they needed to curry favour with their own countrymen, but the town and its surrounding gardens could look London, Kensington Gardens and Hampton Court culturally in the eye.

Hatton (1978); Barlo, Komachi and Queren (2008).

C. Janson, *Plan du Chateau et Jardin Royal à Herrenhausen avec quelques autres Jardins et Maisons de Plaisance à l'Alentour*, Hanover, 1726–7

Manuscript on vellum, 219 x 72 cm

British Library Maps K. Top 50.59-a

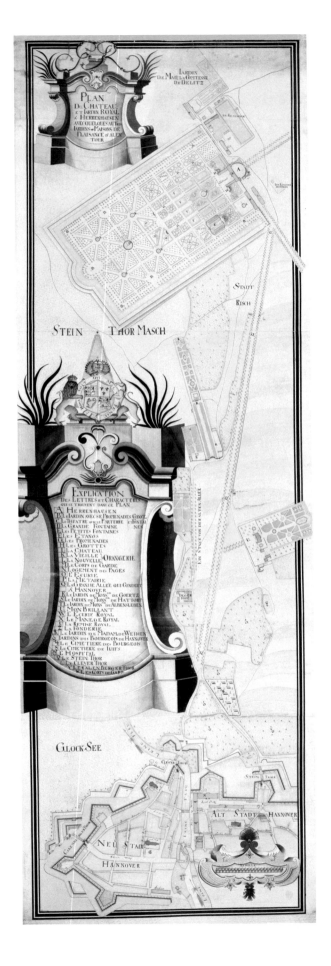

'A vast book of Mapps'

The Klencke Atlas closed

According to the *Guinness Book of Records*, the Klencke Atlas is the largest atlas in the world. It was presented to Charles II on his Restoration in May 1660 by a consortium of Amsterdam merchants, led by Johannes Klencke. The atlas contains forty-one wall maps in thirty-nine sheets from the golden age of Dutch mapmaking, mainly from the workshop of the Blaeu dynasty. The maps cover the whole world and are placed in a binding stamped with the rose of England, the thistle of Scotland, the harp of Ireland and the fleur-de-lys of France (of which Charles still claimed to be king). The atlas is all the more precious today because the maps are in perfect condition, while most other loose examples, subjected to light, dust, damp, heat and the effects of varnish, have perished, or have been left as battered shadows of what they once were.

The atlas was a prestigious gift and the merchants hoped it would win them appropriate rewards. Johannes Klencke himself received a knighthood, but we do not know whether he or his fellow merchants won the commercial privileges that they doubtless hoped for, at a time when English ships were beginning to challenge the supremacy that the Dutch mercantile fleets had hitherto enjoyed on the high seas.

It would be wrong to see the Klencke Atlas simply as an oversized bribe, however. It had two, slightly smaller, fellows, of which one (now in Berlin) had been presented to the elector of Brandenburg and the other (now in Rostock) had been purchased by the last independent Duke of Pomerania. The inspiration behind all three seems to have been Johan Maurits, Duke of Nassau-Siegen, a man of broad cultural interests who was also responsible for commissioning the magnificent 'Marcgraf' map of Brazil (p.43). An example of this is also to be found in the Klencke Atlas.

Johan Maurits may in turn have been influenced, directly or indirectly, by the writings of the Italian theorist Alessandro Citolino. In his book *Tipocosmia* of 1561 Citolino had summoned up the image of an enormous book (*'un libro di estrema grandezza'*) containing maps of heavens and of the world and its parts, as well as images and diagrams which would collectively have contained the world's knowledge. The obvious location for such an encyclopaedic work would have been a princely cabinet. This, as well as its sheer size, may help to explain why in November 1660 the diarist and polymath John Evelyn saw this 'vast book of Mapps in a volume of neere 4 yards large' in Charles II's 'Cabinet and Closset of rarities' in Whitehall Palace, alongside:

> ...rare miniatures of Peter Oliver after Raphael, Titian & other masters ... that large piece of the Dutchesse of Lennox, don in Enamaile by Petito [Petitot]; & a vast number of Achates, Onyxes, & Intaglios, especialy a Medalion of Caesar, as broad as my hand: likewise rare Cabinetts of Pietra Commessa: A Landskip of Needleworke, formerly presented by the Dutch to K. Char: I. ...: a curious Ship modell, & amongst the Clocks, one, that shewed the rising & setting of the son in the Zodiaque, the Sunn, represented in a face & raies of Gold, upon an azure skie, observing the diurnal & annual motion, rising & setting behind a landscap of hills, very divertisant, the Work of our famous Fromantel, & severall other rarities in this royal Cimelium.

Fiorani (2005), p.90; van den Boogart (1979); Whitehead and Boeseman, (1989); de la Bédoyère (1994), p.127.

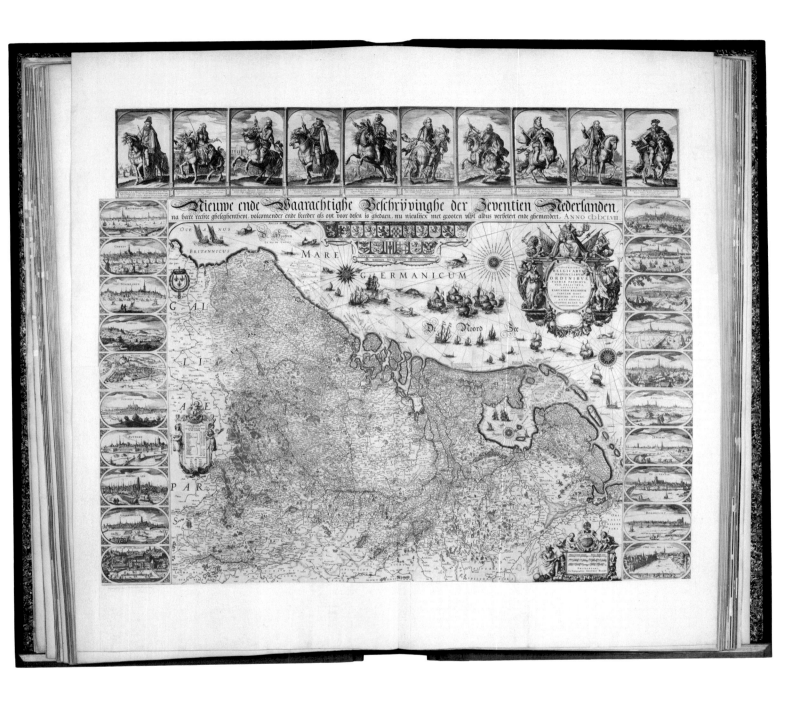

The Klencke Atlas, Amsterdam,
Johan Maurits of Nassau, Joan Blaeu

Atlas with forty-one printed wall maps,
176 x 231 cm (open)

British Library Maps K.A.R.

Atlas open at *Nieuwe Ende Waarachtighe
Beschrijvinghe der Zeventien Nederlanden*
by Joan Blaeu, 1658

The world in your hand: Globes

For a ruler, the ownership of an exotic artefact would enable him or her to display his worldly knowledge, as well as an appreciation of different cultures and systems of understanding. This is the earliest known Chinese terrestrial globe, the work of Jesuit missionaries in China and an example of European cartography originally made for the Chinese Emperor. The Jesuits, missionaries with a remit to promulgate Christian and Western knowledge, had been present in China since the late sixteenth century. By the early 1600s they had become influential figures in the courts of the Ming Rulers. The creators of this globe are thought to be Manuel Dias (1574–1659), who introduced the telescope to China, and Nicolo Longobardi (1565–1655), who had become superior general to the China mission in 1610. Both were respected scholars, and the globe's depiction of the coasts of Africa and Europe would have contrasted with the sino-centric nature of contemporary Chinese cartography. In its treatment of eclipses and meridians and its information about magnetic inclination, however, the globe draws on ideas which had been developed in China far earlier than in the West. The globe was believed to have been taken by British soldiers from an imperial palace in or near Beijing in 1860, and was long cited in the West as a forerunner of the 'modernization' – meaning Westernization – of Chinese mapping. In fact it seems that Western mapmaking concepts were little regarded beyond the Chinese court until well into the twentieth century.

If globes do not seem to have been produced in seventeenth-century China, there flourished in Europe, and particularly in the Low Countries, a strong and highly competitive globe-making industry. Globes were constructed in a variety of sizes using printed globe maps, or gores as they were properly termed, by Joan Blaeu, Jodocus Hondius and Florent van Langeren. Large display globes (pp.60–1) functioned as overt statements of prestige and power. Yet smaller-sized globes could impart equally powerful, but more contemplative, messages to their owner.

The pocket globe, that is a miniature portable sphere made from wood or plaster with a protective case, originated in the 1670s and has been labelled an 'English phenomenon' due to its almost exclusive creation and use there. It was designed to fit snugly into its custom-made case, usually in a durable material such as cloth or snakeskin and often with celestial globe gores pasted to the inside. The whole object would have acted as a faithful reproduction of both the terrestrial and celestial worlds.

Such miniature globes were used for study, amusement and intellectual stimulation in the same way as larger table globes, yet they could also be taken and used on journeys. Because of the skill and – in a word much-used by contemporaries – 'ingenuity' needed to create them, they also became prized possessions, to be stored with freaks of nature and objects of virtu in cabinets. The experience of holding a replica of the world in one's hands would have encouraged a ruler or nobleman to meditate upon his or her own duty towards the world, and on the exalted place he or she held within it. This symbolic function of the globe bears clear similarities with the role of portable orb or *globus cruciger* – hand-held attributes of authority of Christian rulers from the Middle Ages – while also recalling instances of medieval monarchs appearing in public in front of world maps.

The level of technical skill involved in the construction of pocket globes was such that their creators were often also makers of time-keeping devices, navigational instruments and mechanical objects. In 1613 the Duke of Saxe-Weimar reported that at Whitehall he had seen 'a beautiful celestial globe of brass, which when wound up went round of itself. Presented by the Emperor [probably Rudolf II to the King.' A similar gilt clockwork celestial globe, also commissioned by Emperor Rudolf II, is now owned by the Victoria and Albert Museum in London.

Nelson and O'Donoghue (1974), no.C8; Yee (1994), pp.170–202; Dekker and van der Krogt (1993), p.109.

OPPOSITE: *Chinese terrestrial globe*,
Manuel Dias and Nicolo Longobardi,
Beijing, 1623
59 cm diameter
British Library Maps G.35

ABOVE: *Pocket Globe*, J. Moxon,
London, 1679
7 cm diameter
British Library Maps C.4.a.4.(7.)

A Correct Globe with ye Trade Winds,
Herman Moll, London, 1710
7 cm diameter
British Library Maps C.4.a.4.(1.)

*Cary's Pocket Globe, Agreeable
to the Latest Discoveries*,
J. Cary, London, 1791
8 cm diameter
British Library Maps C.4.a.5.(1.)

*Newton and Berry's New Terrestrial
Globe*, London, 1831
4 cm diameter
British Library Maps C.4.a.5.(5.)

Miniaturization: Coins and medals

Some of the most common forms of miniaturization were to be found on coins and medals. They were avidly collected by monarchs, princes and courtiers between about 1420 and 1800, with the coin cabinets being housed in libraries, studies and cabinets. Several coins as well as medals depicted globes, maps and town views, serving in most cases the same function as their giant counterparts. Civic pride could be expressed through a miniature town view, and the doctrinal reluctance to display religious images meant that many German and Swiss city states substituted a town view for a depiction of a patron saint on their coinage, particularly after 1580. Depictions could range from the minute, as on a Nuremberg kreutzer of 1773, to the relatively substantial on a medal.

Commemoration was also an important function. Rulers frequently commissioned medals comparing their burdens to those of Atlas. As early as around 1580, Michael Mercator – grandson of the more famous Gerardus Mercator, map publisher and popularizer of the projection named after him – produced a medal with a dual hemisphere world map on either side to commemorate Francis Drake's circumnavigation of the globe. Commemoration was also one of the purposes behind the production between 1570 and 1670 in the infant Dutch Republic of a long series of silver medals and copper tokens to mark their military successes against former Spanish overlords. In an expression of what the art historian Svetlana Alpers has called the 'mapping impulse' in Dutch art, these medals often contained miniaturized maps and views of the sieges.

The larger, grander medals were sold and distributed to the wealthy at home and abroad, while the smaller copper pieces circulated widely among the people as gambling and reckoning tokens, promoting propaganda for the Dutch as they did so. During the eighteenth century the great powers continued to commemorate famous victories and diplomatic triumphs through cartographic medals. In 1971 the Royal Geographical Society commissioned an accurate medallic map of the moon to commemorate the first moon landing in 1969.

Sometimes the cartographic medals that circulated in royal and aristocratic circles in the late seventeenth century were satirical. Louis XIV of France, the self-proclaimed 'Sun King', set the ball rolling in 1662 with a design that compared him to the sun dominating the world. In 1670 Charles II responded with a medal containing an accurate representation of the world, plus the inscription that Britons were now spread across the globe following the cession of Bombay and Tangiers in addition to existing North American colonies. In 1673 George Bower bettered this on a medal commemorating the ennoblement of one of Charles's mistresses. It featured a picture of Cupid on top of the world – one of the first maps to show the west coast of Australia – and the message that Love or Cupid was the ultimate conqueror of the world! The exiled Queen Christina of Sweden, however, who had abdicated and retired to Rome after converting to Catholicism, would have none of it. She issued medals with a globe surrounded by the message that the world (ie worldliness) was not sufficient for her. The final riposte in this debate came a few decades later as France faced invasion during the War of the Spanish Succession. Moved by the occasion, a German medallist struck a medal showing the globe of the Earth, representing Louis's enemies, eclipsing the sun.

E. Hawkins, A. Franks and H. Grueber (1885); G. van Loon (1732–7); Barber, (1999), pp.53–80; Kollnig, (1987).

A   Zurich, 1/2 taler, 1721, silver, diameter 3.2 cm

B   Nuremberg, 1 kreutzer, 1773, silvered bronze, diameter 1.5 cm

C   M. Mercator, Drake's circumnavigation of globe, 1580, silver (electrotype), 7 cm
    Medallic Illustrations i, p.131.83

D   Capture of Sluys by and loss of Ostende to Dutch, 1604, silver, diameter 5.6 cm
    Van Loon, ii, p.30

E   P. Vestner, Accession of George I, 1714, white metal, diameter 4.3 cm
    MI ii, p.422, 5

F   Royal Geographical Society, Manned landing on the Moon, 1971, silver, diameter 6.5 cm

G   J. Warin, Louis XIV, 1663, silver, diameter 5.45 cm

H   J. Roettiers, British colonization, 1670, silver, diameter 4 cm
    MI i, p.546, 203

I   G. Bower, Ennoblement of Louise de Querouaille, 1673, silver, diameter 2.5 cm
    MI i, p.554, 215

J   G. Hamerani, Christina of Sweden, 1680, bronze, diameter 2.5 cm

K   M. Brunner, Loss of Douai to France, 1710, white metal, diameter 4.5 cm
    MI ii, p.371, 216

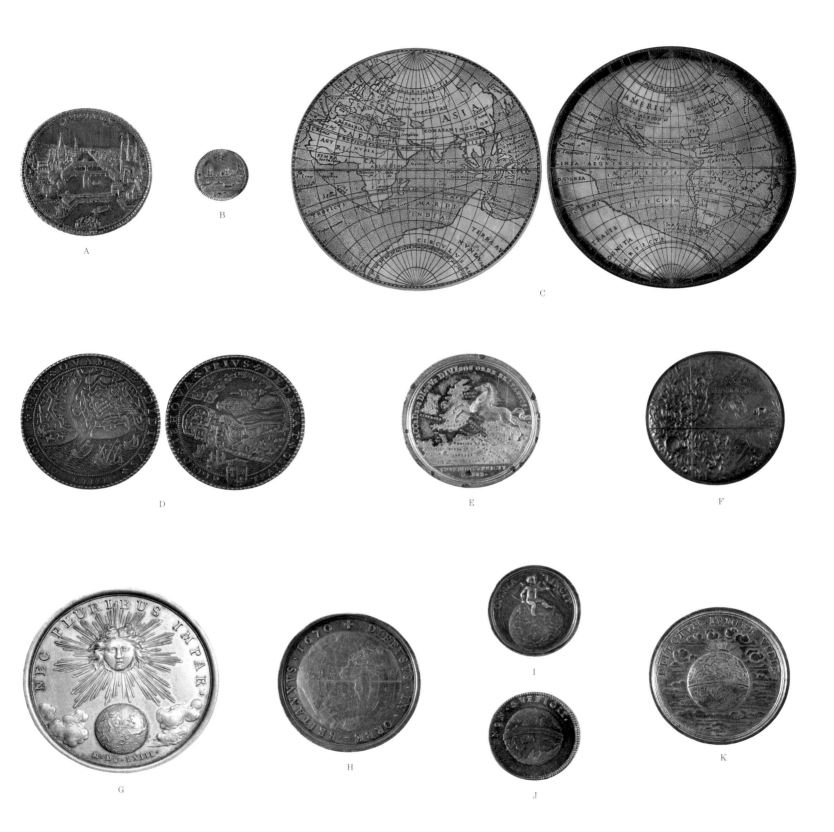

A

B

C

D

E

F

G

H

I

J

K

A free-standing atlas of Europe

George Willdey was a flamboyant London shopkeeper and self-publicist who boasted that in his premises on Ludgate Hill visitors could buy 'the greatest Choice of curious Things in Gold … fine plate, silver, all sorts of Jewels … fine china … the very best Spectacles, Reading-Glasses, Telescopes, Microscopes and Perspective Glasses of all which, and abundance of other curious and valuable Things, no Person hath better Choice, or sells more reasonable'. He also sold maps. In 1721 he added a note on one of his large world maps that 'round the same is added instead of useless ornament a set of twenty maps of the principal kingdoms and states of Europe with particular historical explanations'. The total size of the maps was 'six Feet high and eight Feet four Inches long' and, Willdey pointed out, they would make 'useful and entertaining Screens for large Rooms, Halls and Counting houses'. In a sense they were, albeit less grandly and in printed form, equivalents of the maps that Egnazio Danti had created for the fronts of cupboards in the *Guardaroba Nuova* of the Palazzo Vecchio in Florence 150 years earlier.

After Willdey's death the copperplates passed to his son, and from him to Thomas Jefferys, geographer to the king, and one of the ablest and most prolific mapmakers of the mid-eighteenth century. He updated the world map, for instance, by showing the course of Admiral Anson's circumnavigation of the globe in 1743, pillaging Spanish ships as he went. It was Jefferys who sold the set of maps on this screen.

The complete set covers the whole world, with two maps of the heavens in the corners. The continents distant from Europe were catered for in the world map, and there are twelve sheet maps of different parts of the European mainland – including one each for England, Scotland, Ireland, Oxford and Cambridge, and the area twenty miles around London. The balance between Britain and the world very much reflected English perceptions at the time, just as the bias towards Italy in the painted galleries of the late sixteenth-century Italy reflected theirs.

These map screens seem to have been popular at the time, and they were mentioned in Sheridan's comedy, *School for Scandal*, as late as 1777. They would have given a flattering impression of the extent of its owner's knowledge and interests, and would have provided an informal geographical education for family, visitors and servants alike. Individual maps could be cut up for mounting atop poles as screens to protect complexions from from the heat and flames of a fire. Two such small map screens were purchased for Ham House in Middlesex in 1743 (p.82), and they are still to be seen in the library there. Time has taken its toll on these fragile objects, however. Although single sheet maps occasionally come onto the market, only one other large screen of this type seems to have survived.

The National Art Collections Fund, 1997 *Review*, p.84; Hodson (1984), i, pp.142, 182.

George Willdey and Thomas Jefferys, *Four-fold screen, with twenty-one hand-coloured copper-engraved maps*, London, *c*.1750

Canvas with a pine frame, 180 x 256 cm

British Library Maps Screen 2
Purchased with assistance from the Art Fund

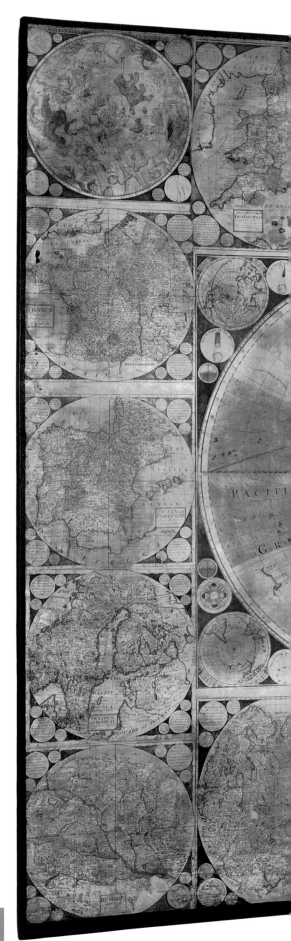

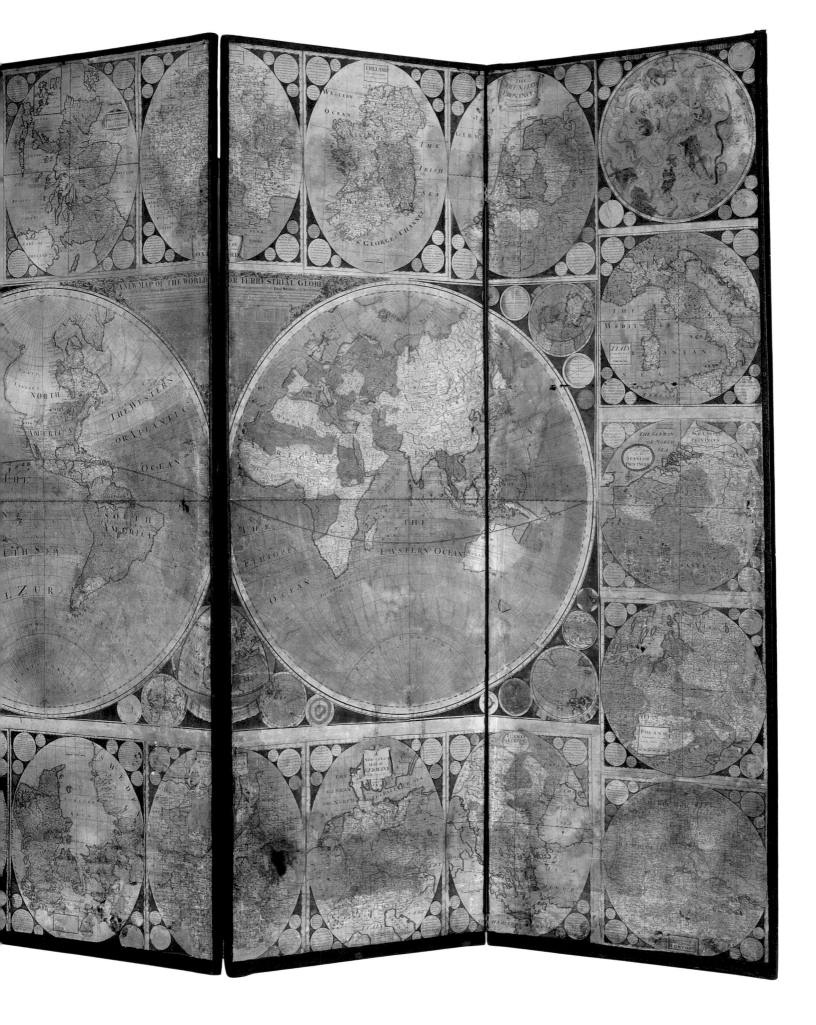

The refectory of the monastery of San Lorenzo, Naples

THE SECRETARY OF STATE'S ROOM

In 1338, a few years before creating the great world map in the *Sala del Map-pamondo* in the Palazzo Pubblico in Siena, Ambrogio Lorenzetti painted two great murals, allegorically representing good and bad government, in the adjacent room – used for meetings by the magistrates who ruled the city-republic. The backdrops to the allegorical scenes were Sienese land and cityscapes, and they were depicted with a realism that verged on the cartographic. Over the following centuries, this example was copied by the leaders of several Italian communes, republics and principalities. They proceeded to decorate the walls of their council chambers and rooms where the ruling elites determined policy with maps and landscape and townscape panoramas.

As modern bureaucratic forms of government gradually developed from the sixteenth century, and the role of the 'secretary' (the predecessor of the government minister) evolved, so maps began to appear on his walls too. They also continued to adorn the walls of government committees (for instance, those of the seventeenth-century Committee of Plantations, the ancestor of the Colonial Office of Victorian England),[85] and councils, as well as the courts of governors of trading companies – notably the British and Dutch East India Companies. Before 1700 the meeting rooms of councils and the offices of secretaries were mainly to be found within the walls of palaces that mushroomed in size in Baroque Europe. In certain countries, notably France, this practice continued throughout the eighteenth century, but elsewhere government offices gradually became established beyond the palaces, as the headquarters and offices of the trading companies always had been.

The maps to be found on the walls of such rooms fell into two broad categories. Some were there to act as symbols of the state, simplified and censored when necessary to avoid betraying sensitive information to potential enemies. This was what happened particularly in larger council chambers where diets, estates and parliaments met. Sometimes mistakes happened, as in Venice when the Doge's Palace had to be redecorated following disastrous fires in 1574 and 1577. In 1578 Cristoforo Sorte, an artist and engineer as well as a mapmaker, was ordered to prepare a vast and extremely detailed painted map of the Venetian Terraferma. The map, measuring no less than 12 metres in length by just over 4 metres in height, was to fill the internal wall of the Senate Chamber in Venice. Over the following years Sorte undertook a detailed survey of the territories involved. The authorities, however, had second thoughts about the large painting on security grounds, and in 1582 asked Sorte instead to prepare a smaller and less detailed version measuring less than 2 metres in height and length for public view. The separate, more detailed maps of the Venetian provinces were kept under lock and key in a neighbouring room.[86]

Luigi Rodriguez, the artist responsible for the small frescoed maps of the dominions of the kingdom of Naples – created in the 1590s to adorn the lunettes of the vaults of the council chamber where the parliament of Naples met (now the refectory of the monastery of San Lorenzo Maggiore in Naples) – avoided such problems by copying, in much simplified form, Nicola Antonio Stigliola's manuscript maps of the provinces of the kingdom that had been surveyed in the previous decade. The mural maps contained only information that the Spanish authorities felt could be safely divulged.[87]

In England it was a slightly different story. In 1596 Elizabeth I's Lord High Admiral, Charles Howard, Lord Howard of Effingham, commissioned weavers in Delft to create tapestries reproducing the series of strategically bland maps illustrating the course of the Spanish Armada (created in 1590 by Robert Adams and engraved by Augustine Ryther). The tapestries were duly completed and eventually hung on the walls of the House of Lords. There was nothing confidential about the content, and their images and luxuriousness made them a great attraction for visitors to Westminster, while encouraging British legislators with visions of a heroic past. The tapestries were destroyed by the fire of 1834 which devastated the whole of the Palace of Westminster,

Lunette with simplified map in the refectory of the monastery of San Lorenzo

Anon, Meeting of the directors of the Dutch East India Company in a map-lined office, Amsterdam, 1768

but luckily they had been engraved by John Pine, the finest engraver of his day, in 1739.[88]

The second sort of map, typical of a minister's office, was the working map, often hung from the wall, but which could be taken down to be studied at a table, even though some of them could be sizeable. Such maps had been created in northern Italy, and particularly in Venice and in the duchy of Milan from the 1290s.[89] The majority were utilitarian and objects of no particular beauty; their twentieth-century descendants can be seen in the Cabinet War Rooms in Whitehall. A significant percentage, however, such as a handsome map of the surroundings of Verona and Lake Garda, created in about 1450 and now in the State Archives in Venice,[90] were executed much more carefully. These maps were finished with a degree of finesse and artistry, and at far greater expense than was required for them to perform their essential function.

The reasons for this were many and varied. Clarity in lettering, and well selected and balanced colouring, often served to increase the utility of a

François-Hubert Drouais, *The Comte de Vaudreuil*, 1758. National Gallery, London

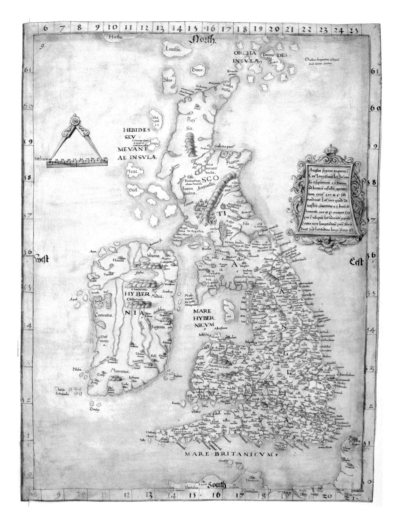

Maurice Griffith (?), 'Cottonian' map of the British Isles, *c*.1537.
British Library Cotton MS Aug.I.i.9

map. In the case of the printed maps, created as a commercial speculation (as with the numerous 'Theatre of War' maps published from 1695 and intended to enable purchasers to follow the course of a campaign), flamboyant decoration was probably an attempt to conceal the re-use of old regional maps, reduced to the same scale and with information about the war and its battles added.[91] In other instances the person responsible for commissioning a manuscript map (or a printed map intended for restricted circulation) may have seen in it the means of making a good impression on an important person. This was probably so with Maurice Griffith, Archdeacon of Rochester, who presented a handsome manuscript map of the British Isles to Henry VIII as a New Year's gift in 1538.[92] In other cases, the mapmaker himself saw a career opportunity, or wanted to make the point that he was more than a mere draughtsman. In the course of the eighteenth and nineteenth centuries professional artists of the calibre of Paul Sandby were employed as drawing masters at military academies[93], and numerous military draughtsmen dreamt of making careers as artists. A few, such as William Payne, even succeeded.[94]

85   Jeanette Black, *Commentary on The Blathwayt Atlas. A collection of forty-eight manuscript and printed Maps of the seventeenth century relating to the British Overseas Empire in that era, brought together about 1683 for the use of the Lords of Trade and Plantations by William Blathwayt, Secretary* (Providence, Brown University, 1975).

86   Juergen Schulz, '*Cristofor Sorte e il Palazzo Ducale di Venezia*', and '*Nuove mappe e disegni di paesaggio di Cristoforo Sorte*' in *La Cartografia tra scienza e arte*, pp.97–101, 113–39; Emanuela Casti, 'State, Cartography and Territory in Renaissance Veneto and Lombardy' in *History of Cartography* iii, pp.902–4. The simplified map of the Terraferma is now lost, but so accurate were these maps that following the fall of the Republic of Venice nearly 200 years later they were carted off to Vienna for the use of the new Austrian rulers. Some remain in Vienna.

87   Vladimiro Valerio, *Società Uomini e Istituzioni Cartografiche nel Mezzogiorno d'Italia* (Florence, Instituto Geografico Militare, 1993), pp.54–6. I am most grateful to Vladimiro Valerio for this information and for his great assistance in obtaining images of the refectory and its frescoes; Fiorani, 2007, p.813.

88   *Lord Howard of Effingham and the Spanish Armada. With exact facsimiles of the 'Tables of Augustine Ryther' AD 1590 [i.e. the plates in 'A Discourse concerninge the Spanishe fleete inuadinge Englande' by P. Ubaldini, engraved by A. Ryther after Robert Adams], and 'The Engravings of the Hangings of the House of Lords' by John Pine [after Cornelius Vroom], AD 1739. With an introduction by Henry Yates Thompson* (London, Roxburghe Club, 1919). I am grateful to Hilary Turner for background information on the commissioning and the location of the workshop that produced the tapestries.

89   See P.D.A. Harvey, 'Local and Regional Cartography in medieval Europe' in *History of Cartography* i, pp.478–82.

90   A detail is reproduced in colour in P.D.A. Harvey, *The History of Topographical Maps* (London, Thames & Hudson, 1980), plate IV.

91   Dirk de Vries, 'Dutch Cartography' in Robert P. Maccubbin & Martha Hamilton-Phillips (eds), *The Age of William III & Mary II. Power, Politics and Patronage 1688–1702* (Williamsburg, College of William and Mary, 1989), pp.109–110; Marco van Egmond, *Covens & Mortier. A Map Publishing House in Amsterdam 1685–1866* (Houten, Hes & de Graaf, 2009), pp.150–1.

92   Peter Barber, *King Henry's Map of the British Isles* (London, Folio Society, 2009), particularly pp.94–101.

93   Jessica Christian, 'Paul Sandby and the Military Survey of Scotland' and Nicholas Alfrey 'Landscape and the Ordnance Survey, 1795–1820' in *Mapping the Landscape. Essays on art and cartography*, eds. Nicholas Alfrey and Stephen Daniels (Nottingham, Castle Museum, 1990), pp.18–22, 23–7; John Bonehill and Stephen Daniels (eds), *Paul Sandby. Picturing Britain* (London, Royal Academy of Arts, 2010), pp.17–18. As late as the 1820s and 1830s Theodore and Thales Fielding, brothers of the better-known Copley Fielding, were employed as drawing masters at the East India Company's military academy in Addington and the Royal Military Academy at Woolwich, and in 1802 John Constable seriously considered taking employment as a drawing master at the Royal Military Academy (see their respective biographies in the *Oxford Dictionary of National Biography*).

94   Donald Japes, *William Payne – a Plymouth experience* (Exeter, Royal Albert Memorial Museum, 1992).

Security

In 1601 a Spanish force landed off Kinsale near Cork in southern Ireland. Although the invaders were speedily repelled, the event had a traumatic effect on the English authorities, obsessed for the next two centuries by fear of a repeat invasion. A new castle, named James Fort in honour of James I, was speedily constructed. In the mid-1670s, at a time of widespread fears of a 'Popish' plot aided by France and Spain to overturn the established Protestant governments, the decision was taken to commission William Robinson to rebuild the medieval Ringcurran Castle in accordance with the latest fortification theories.

This map was probably created in a chartmaking workshop on the shores of the Thames, east of the Tower of London, from a detailed working plan sent over from Ireland. Drawn on vellum, it shows the town of Kinsale and its harbour with inset coastal views, and a large plan of 'Ringcoran' castle. Great care has been taken in its execution, and its splendour and use of gold suggests that it was created for presentation to King Charles II himself – perhaps following the completion of the building works in 1677, when royal permission was sought for the fortress to be renamed in his honour.

The British Library owns another example of the map, prepared by the same workshop, but lacking the gold and painted onto linen, and this may have been intended for administrative use.

Charles was an enthusiast for military architecture, as the earlier fortification plans in King George III's geographical collections in the British Library demonstrate, and he would have been both informed and delighted by the map. It would also have made an attractive object for display and propaganda, suggesting to visitors that the English were in full control even of this most vulnerable part of Ireland. But such a lavish gift would also have served to remind the king of the loyalty of its donor – probably his representative in Ireland, the Lord Lieutenant, James Butler, Duke of Ormonde.

In the event, the new fort did not live up to expectations. Forces loyal to the exiled James II seized it in 1689, although John Churchill, Earl of Marlborough, was able to recapture it for William III in the next year. He achieved this with little difficulty by bombarding the fort from the overlooking hills. The fort continued in military use until 1922 and is now a major tourist attraction.

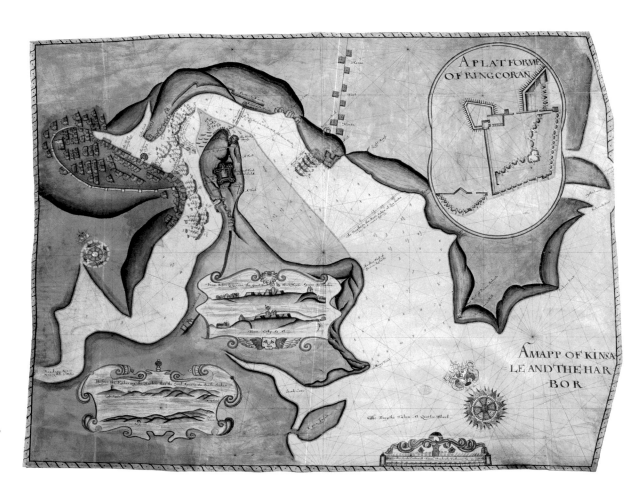

Anon, *A Mapp of Kinsale and the Harbor*, London, c.1677
Manuscript on vellum, 74 x 97 cm
British Library Maps K. Mar. 4.20

Safeguarding the key to England

Dover Harbour was a perennial cause of concern to English rulers during the sixteenth century. Described as 'the key to England' by the thirteenth-century chronicler Matthew Paris, the port was also the key to England's communications with mainland Europe – particularly, before 1558, to Calais, the sole remaining English possession on French soil. Improvements for its defences and harbour facilities were constantly being proposed, making Dover Harbour the most mapped place in the kingdom. But all the while the effort was at risk of being nullified. Shingle (referred to here as 'beach'), driven along the coast by wind and currents, was rapidly filling the

harbour (top right on the plan). The harsh winter of 1551–2 brought the situation to crisis point.

This plan, which has south at the top, was prepared for presentation to the Privy Council on 22 November 1552 by the anxious burgesses of Dover. It shows the inroads made by the shingle in recent decades and illustrates proposals to keep it at bay. The engineers behind the map develop existing proposals for the protection of Dover Harbour by showing two immense piers or moles stretching south from the old town and north, in this case, from Crane Head. They further suggest the construction of additional small jetties to the south of the King's Pier and Archcliff ('Heckclyfe'). The intention was that the piers and new jetties would hold back the shingle, while the current of fresh water from the stream running through the town would help to push it away. The newly proposed jetties are marked by dotted lines, and the big piers portrayed in earlier plans by

John 'Warwick' Smith, *Watercolour View of Dover* (detail), *c*.1790. Private collection

continuous lines. The course of the town stream's current is depicted in purple, to distinguish it from sea water. Pencilled outlines show that the pier to the right (south) was originally intended to be longer and to take a different, partially semicircular, form. The text on the map ends with a request that the Council appoint expert commissioners to visit Dover without delay, to assess the situation and talk to local people 'least [*sic.* = lest] this winter wil be more hurtfull then yet is knowen'.

Most of the map is pictorial and a textbook example of chorography as described by Ptolemy. The image is constructed from multiple viewpoints which have been subtly merged. One viewpoint, towards the castle, is taken from the high ground to the northwest, an angle selected by artists such as John 'Warwick' Smith in the eighteenth century. Another viewpoint, over the bay, is from the castle; a third is from the shingle under the cliffs towards the harbour, while the sea and harbour works are shown in

bird's-eye perspective. The artistic effect is enhanced by the harmonious, although not totally realistic, use of colours, and by the careful depiction of castle, town (with its church), Bight (a huddle of houses by the harbour) and bulwarks. Folkestone, further to the west, is treated in a more generalized way.

The proposals could have been conveyed much more simply, but less elegantly. One feels that the burgesses of Dover would have agreed with the courtier and theorist Sir Thomas Elyot who, in his influential *Boke named the Governour* of 1531, wrote that 'where ... that which is called the grace of the thing is perfectly expressed, that thing more persuadeth and stirreth the beholder' (Everyman edition, 1962, p.24).

Minet (1922), pp.185–225; Macdonald (1937), pp.108–26; Colvin (ed.), *History of the King's Works* (1982), pp.729–78, particularly p.751.

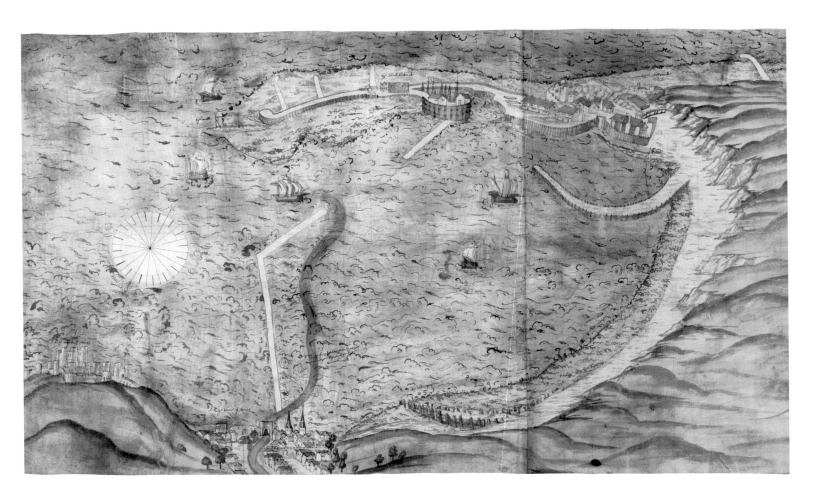

John Rogers and Sir Richard Cotton (?), *Ffor dovour pere. Thys platt exhibited vnto the kinges maiesties [?] most honorable pryvey counsell the xxiith daye of the monthe of november anno regis Edwardi Sexti sexto: de domini 1552,* Dover (?), 1552 (left side only illustrated)

Manuscript on vellum, 65 x 166 cm

British Library Add. MS 69824

Trying to win support from Mr Secretary Cecil

This elongated map depicts the valley of the Trent – then as now, one of England's economic heartlands, whose importance is derived from water power. This may explain why the River Trent had already been mapped, with relative accuracy, by the early fifteenth century on the Gough Map, now in the Bodleian Library in Oxford.

The map is oriented with west at the top. It represents an extensive area, but its primary focus is the area immediately around Newark, shown in great detail. The map seems to have been created in connection with a legal dispute over water rights. In about 1558 a Mr Sutton, lord of the manor of Averham and Kelham, is said to have cut a channel near Farndon, diverting much of the River Trent's water into what had been a small stream flowing through Averham, Kelham and Muskham. The resultant lack of depth in the river not only stopped boats from reaching Newark, but, much more seriously at the time, it left Newark's six mills high and dry. The mills were used both for grinding corn and for the fulling of cloth, Newark's main industry. The mills owners felt compelled to go to court.

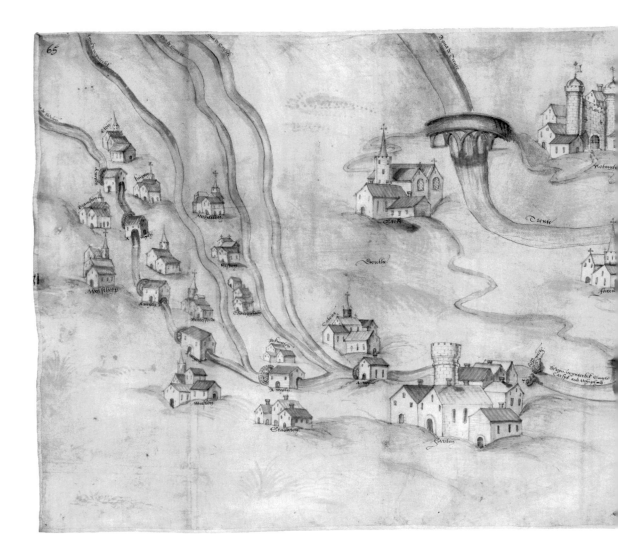

Anon, *Mills on the River Trent*
Nottinghamshire, Nottingham, *c.*1558
Manuscript on vellum, 45 x 122 cm.
British Library Cotton MS Aug.I.i.65

The map may have been commissioned by Mr Sutton to illustrate his position. It shows a 'weir' near Averham, the 'two mylnes of master Sutton' (of which the remains of one can still be seen on the ground) and the 'mylnes of Newarke', all of them richly supplied with water. In the event Mr Sutton was forced to agree to construct and maintain in perpetuity a new weir further upstream beyond Averham. This was placed where the Trent and the cut ('goore') leading to Newark separated, to ensure that an adequate flow of water continued to the mills there.

The map may also have been meant to win the support of the queen's influential secretary Sir William Cecil, later Lord Burghley, who was notorious for his habit of demanding maps to clarify situations. Many of the maps used or commissioned by Cecil are to be found in the Cotton collection in the British Library, from which this map also comes.

The map itself, however, would have looked old-fashioned by 1558. Moreover it includes several features that have nothing to do with its principal focus. Great attention is paid to bridges and fords. Nottingham and villages and rivulets far to the south of Newark are shown, with the rivulets named in Latin and other inscriptions in English. This suggests that the map derives in style as well as content from an earlier one, made to serve a quite different purpose. It is possible that the older map was created as early as 1531–2 in connection with an important act of Parliament providing for the maintenance of bridges and waterways, the cost of which was placed on the relevant but undefined parish. The information on the original map would have been useful in identifying the appropriate parish – hence the naming of the villages – in the years when maps were just beginning to be used for administrative purposes in England.

It is perhaps the elegance, size and archaic style of the map that have prompted the suggestion that it may have been a preparatory drawing for a tapestry. It seems more likely that the maker of the original map, unfamiliar with mapping conventions and desiring to create something visually pleasing for presentation to an important person, drew it to resemble a tapestry.

Salisbury (1983), pp.54–9 (with colour reproduction); Duffy (2001), p.52 ff.

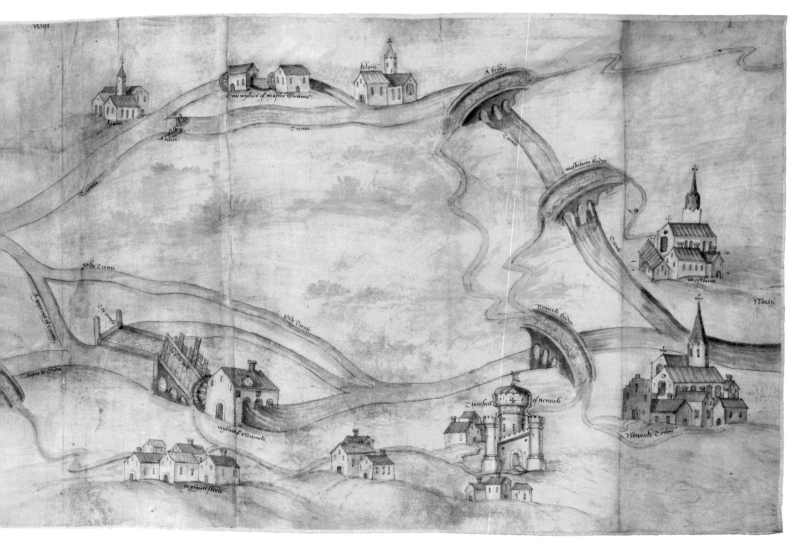

Annotated proof

The use of maps to illustrate boundaries has been one of their main purposes in the arena of government and administration. One example of a dispute between two particular spheres of influence concerns the environs of the southern German town of Nuremberg during the mid-sixteenth century.

The increasing influence of this powerful town over its surroundings had led to a widening of its jurisdiction during the previous century, as the agrarian population gradually came under its authority. Of key importance to Nuremberg was control of the large area of forestland surrounding it, not only providing protection, but also supplying a range of commodities needed to sustain its growth. In addition to the supply of wood, forest quarries supplied building materials, meadows along the bank of the River Pegnitz provided grazing pasture, and the production of honey, wax and garlic provided Nuremberg with a healthy export trade. Until the early fifteenth century the forest had been under the control of the burgraves of Nuremberg – hereditary rulers who were also counts of Hohenzollern and were subsequently to be created margraves of Brandenburg-Ansbach and Brandenburg-Bayreuth. Yet even after wresting authority from these hostile neighbours, the town council needed to exercise tight control to defend the forest against incursions from local villagers and adversaries.

A map of the environs of Nuremberg was published in 1563 by Georg Nöttelein (d.1567), but this example, a pre-publication 'proof' state lacking the engraved lettering, was especially prepared in order to assist the town council in their decision-making. It has been annotated by hand to illustrate various strategic points on the forest's periphery, indicating important farms and hunting lodges used by forest guards in the town's employ. Another map sign marks villages in which outlaws or resistance to town control may have been concentrated. These places occur mainly around the southern and western edge of the forest bordering the Hohenzollern-controlled district of Ansbach, but several lie outside the green-coloured area of Nuremberg control. The identification of villages situated on roads leading into the forest suggests concern not only with encroachment but also the safe passage of goods in and out of Nuremberg.

By highlighting villages on the periphery of their area of jurisdiction, the town council hoped to consolidate their holdings and to anticipate complaints from their neighbours. This map may therefore be seen as an expression of Nuremberg's positional tactics in response to a specific legal dispute. It may even constitute an attempt by the town to bring further land under its control. Nuremberg had gained land to the east of the forest through siding against the Elector Palatine during the Bavarian War of Succession of 1504– 5, and such opportunism benefited from clear recognition of one's strengths and weaknesses.

The final version of the printed map does not include this handwritten information, for such thoughts and fears the town council sensibly wished to keep to themselves. Nöttelein, a mapmaker who was also a church organist, produced a number of maps for the council, and 'proof' copies such as this may have been printed in considerable numbers for future use. Such maps would have served as base maps to be annotated, like this example, and used in disscussion to illuminate particular, confidential problems for Nuremberg's ruling elite as the need arose. Annotations on such maps might occasionally have been added long after the published version of the map had appeared.

Cartographia Bavariae (1988), pp.58–61; Meurer (1991), pp.208–9; Eiden and Irsigler (2000), pp.43–57.

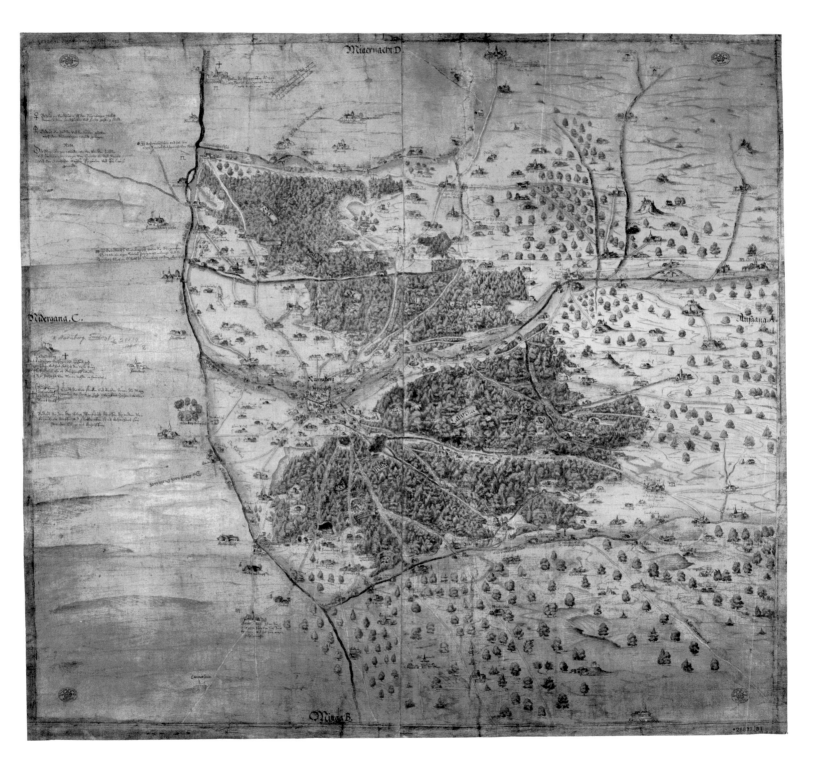

Georg Nöttelein, *Große Wald und Fraischkarte von Nürnberg*, Nuremberg, c.1563
Copperplate engraving on four sheets, 88 x 91 cm
British Library Maps *28839.(8.)

Clemency, patriotism and self-promotion

The aftermath of the defeat at Culloden on 16 April 1746 of the forces of Prince Charles Edward Stuart – representing his father James, the Stuart or 'Jacobite' claimant to the thrones of Great Britain – left John Elphinstone with a dilemma and an opportunity. A member of a prominent Scottish family, he himself was serving in the Hanoverian forces. He had come to official notice since being appointed a practitioner engineer in March 1745 by creating a much-praised map of Scotland that was soon to be printed. However, many of his relatives had been loyal to the Stuarts, and he seems to have been proud to be Scottish at a time when many English people regarded the Scots collectively with deep suspicion, fear and loathing. With a skill in depicting figures unusual among military draughtsman, Elphinstone may have felt that given the proper patronage he had a possible future as an artist.

All of these strands seem to come together in this map, which Elphinstone was asked to produce in the autumn of 1746 for the Earl of Albemarle, who succeeded the Duke of Cumberland as commander-in-chief. At a superficial level, this is just another version of his earlier map of Scotland, to be used to pursue the Hanoverian policy of pacifying the Highland clans. It emphasizes the roads and forts created over the previous decades and, rather less successfully, the mountains. On another level, however, it is an attempt to remind the Hanoverian authorities – and especially Albemarle, who particularly disliked the Scots – that not all Highlanders were Jacobites, and that some Jacobite Highlanders had noble qualities, even while underlining his personal loyalty by celebrating recent Hanoverian successes and satirizing the most disreputable of the Jacobites. This challenging combination was conveyed through an accomplished display of his qualities as a mapmaker and as an artist.

On the map Elphinstone reproduces Thomas Sandby's plan of the Battle of Culloden, now in Windsor Castle, and he marks the sites of Culloden and of another decisive Hanoverian victory over the Jacobites at Glenshiel. The top-left of the map is filled with views of the Scottish strongholds that had never fallen to the Jacobites: Edinburgh, Stirling, Blackness and Dumbarton. The map's title panel is adorned with illustrations of surveying instruments, hinting at the accuracy of Elphinstone's survey.

The rest of the map illustrates Elphinstone's other objectives. At the top-right is Simon Fraser, Lord Lovat, while on trial for his life in London. He is shown sitting under a table listing the clans that had supported the Hanoverians and the Jacobites respectively. Copied from William Hogarth's famous print, published in August 1746, this shows Lovat looking as rascally as his reputation suggested. At the right side of the table, however, is a much more noble kilted Highlander, also in chains, recalling the Jacobite commander Lord George Murray whose probity was recognized by both sides. At the bottom-left, by the scale bar, is another heroic Highlander lamenting a slain comrade.

Thus in addition to providing the Earl of Albemarle with the factual information that he required as commander-in-chief, Elphinstone sends a subtler plea for clemency for his countrymen and patronage for himself.

Anderson (2010).

John Elphinstone, 'A New Map of North Britain done by order of the Right Honourable the Earl of Albemarle, Commander-in-chief of His Majesty's Forces in Scotland', Edinburgh (?), 1746
Manuscript, 146 x 108 cm
British Library Maps K. Top 48.22

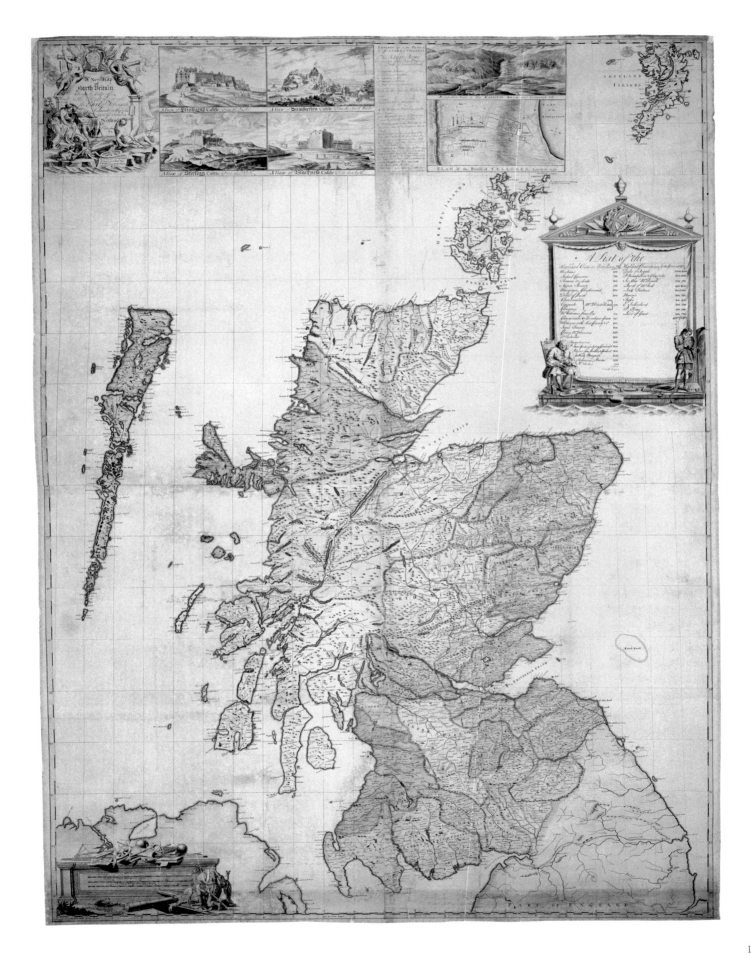

111

The art of war

THE BOY [HENRY PELHAM] AND THE FLYING SQUIRREL. (FROM THE ORIGINAL PAINTING BY COPLEY, NOW IN POSSESSION OF MRS. JAMES S. AMORY.)

John Richard Clayton, *Boy With Squirrel*, 1860, after an original painting by John Singleton Copley. British Museum, Dept Prints and Drawings, 1894, 0102, 68

The circumstances surrounding this map of Boston, created in 1777 by Henry Pelham (1749–1806), are closely tied to the opening events of the American Revolutionary War and the siege of the British at Boston by American forces following the Battle of Bunker Hill on 17 March 1775. The loyalist Pelham, though not a military man, was given privileged access to the newly built defences surrounding the city, as well as existing military maps, for the completion of his map, dedicated to the then British secretary of state, Lord George Germain. Supplementary to its function as a military map, indicating lines of fire and of communication, harbour depths, batteries and defences, the map informs us of the kind of artistic endeavour which has accompanied the depiction of battles and wars on maps. It also reflects technical developments within the London artistic community during the later eighteenth century.

It might seem unnecessary and inappropriate for a military map to have been created using what was then still a moderately new artistic technique of aquatint. This variation of copper engraving allowed for more subtle, tonal effects, with more similarities to watercolour wash than straightforward line engraving. However, the time frame offers the answer here. While Pelham did indeed prepare the map during his time in Boston during the siege of 1775 and 1776, it was eventually published in London just under a year after the British retreat from Boston, preceding Pelham's arrival in the capital a few months later. Thus although Pelham's preparatory material may have been of some use during the siege, the published map was meant for an interested audience in Britain. This potentially included the secretary of state, who would have appreciated a detailed map of one of the largest cities in North America for reference purposes. The reproduction of Pelham's surveyor's pass in the top-left portion of the map would have underlined its reliability in the face of competition from numerous other maps of Boston produced at the same time. This, together with its sublime appearance, would have appealed to the prospective purchaser.

The artist responsible for engraving the map was the aquatint engraver Francis Jukes (1745–1812), who had previously worked with the artist Paul Sandby (1731–1809). Pelham may well have met Jukes during a brief stay in London before Pelham moved to Ireland later in 1778. It is not difficult to explain Pelham's meeting with Jukes and his interest in new artistic techniques: his father Peter Pelham had been one of the earliest British artists to experiment with the engraving technique of mezzotint before emigrating to America in around 1727. Soon after his arrival in the colonies, the elder Pelham had married the mother of John Singleton Copley (1738–1815), the well-known portrait and history painter. Henry appears to have been especially close to his half-brother, since he is identified as the young man in Copley's *Boy with a Squirrel* of 1765.

With such a creative background it is strange that Pelham did not engrave the printed map himself. Trained as a draughtsman, he worked as a painter of portrait miniatures, and his surviving letters include sketches made of the defences at Bunker Hill. Pelham also exhibited work at the Royal Academy of Arts in London in 1777–8. It is more likely, however, that the skill of engraving in aquatint was beyond his ability, and that expenditure on an expert such as Jukes would have been vindicated by the revenue gained through the map's sale. Pelham's decision to publish the map in England, a full year after the siege had ended, was justifiable in a commemorative context, while his unusual use of aquatint may have softened the image, if not the impact, of British defeat.

Krieger, Cobb and Turner (1999), pp.184–5; Archer (2006), pp.183–97.

Henry Pelham, *A Plan of Boston in New England with its Environs...*, London, 1777

Aquatint engraving on two sheets, 107 x 71 cm

British Library Maps *73430.(5.)

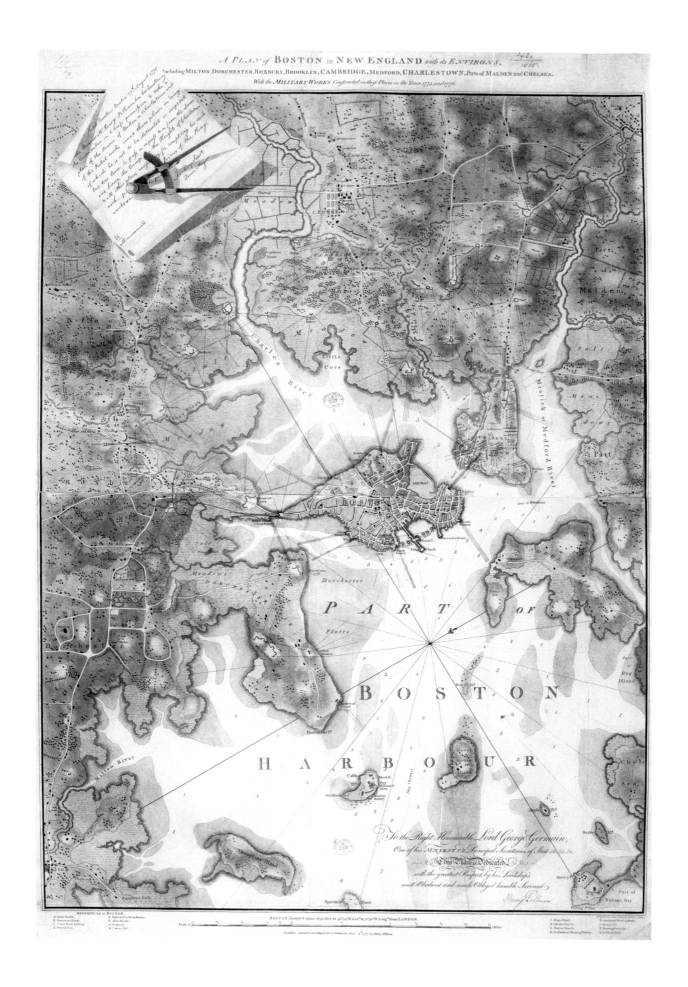

A PLAN of BOSTON in NEW ENGLAND with its ENVIRONS.

Including MILTON, DORCHESTER, ROXBURY, BROOKLIN, CAMBRIDGE, MEDFORD, CHARLESTOWN, Parts of MALDEN and CHELSEA.

With the MILITARY WORKS Constructed in those Places in the Years 1775. and 1776.

113

The finished product

The aesthetic qualities of manuscript military maps did not detract from their function as working documents from which important decisions were made. Rather, precision – together with the careful rendering of topographical features and naturalistic range of colours – was essential for the map to replicate a piece of land for ministers and high-ranking military and naval commanders, usually distant from the areas depicted and requiring an understanding of the situation. Military maps also offered the surveyor-artist the chance to exhibit his skill in front of an influential audience.

Display maps, such as Captain Mark Wood's map of the Hughli River of 1785, were the end results of working surveys. These would have generated a large number of notes, sketches and annotated plans, but their final audience was necessarily limited. The artistic skill of British military surveyors, coached by professional artists such as Paul Sandby (1731–1809) at the Royal Military Academy at Woolwich, was as necessary a quality as the accuracy of their work in the field. Such skill became a trademark of British military surveying in the arenas of northeastern America in the 1760s and 1770s, and of the work of the first Ordnance Surveyors in England and Wales towards the end of the eighteenth century.

Through a series of (partly opportunistic) interventions, the British East India Company had assumed control of the wealthy Indian province of Bengal by around 1770. This fresh phase in British involvement in India was to see a much greater level of administrative involvement than had been seen previously with the Company, whose corruption was being investigated by the government by the late 1770s. The area of the Hughli River immediately downstream from the Company's headquarters in Kolkata was seen as particularly strategically important. Wood's map shows both the newly rebuilt Fort William defending the river and the lines of communication between it and forts on the west bank, and further to the south. To General Sloper, the dedicatee, the map would have brought the district to life – and it seems to have had results. Captain Wood was promoted to the post of surveyor general the following year.

Further possibilities for the district would have been visible in the map to the officials who looked upon it. After 1777 the East India Company resumed control of the collection of rents from local landowners, formulated projects for reforming the land tenure system and agrarian farming, and identified drainage projects and land purchases, aimed at maximizing profits. The statement that the map was *originally intended for military use* gains added significance in the context of an increased British administration of the land so faithfully captured on paper by Captain Wood.

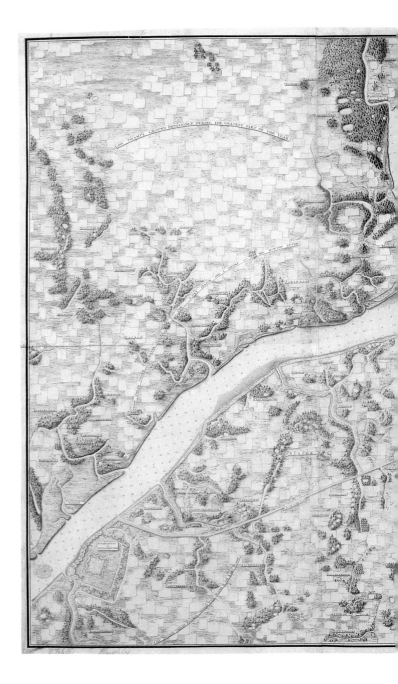

Mark Wood, *Survey of the country on the eastern bank of the Hughly, from Calcutta to the Fortifications at Budgebudge, including Fort William and the Post at Manicolly Point...*, Kolkata, 1785
Manuscript, 74 x 140 cm
British Library Maps K.Top 115.38

Cartographic coaxing

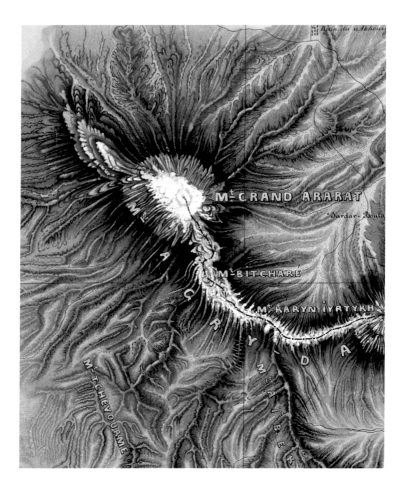

This handsome, hand-drawn map at a scale of 1:73500 (a centimetre to about three-quarters of a kilometre or about an inch to one mile) was presented to the Persian government in 1870 by representatives of the United Kingdom and Russia. The signatures of Sir Henry Elliot and General Nikolai Pavlovitch, Count Ignatiev, the British and Russian ambassadors to the Turkish Sultan near the bottom edge certify the map's official status. It is a fine example of the Russian style of military mapping in the middle to late nineteenth century, combining strongly coloured but sensitively differentiated depictions of the terrain with clear lettering. The dotted red line running across it shows the proposed line of the frontier between Persia and the Ottoman Empire.

The map forms part of a chapter in one of the longest continuous efforts to plot a border anywhere in the world. The approximate line of the border had first been described in May 1639 in the Ottoman–Persian treaty of Zuhab, the oldest explicit demarcation agreement based on geography rather than on purely the clarification of judicial rights between states. Unfortunately the wording of the treaty was vague, and the reality on the ground was a wide and porous autonomous zone that tribesmen crossed at will, despite the existence of border posts scattered along a nominal frontier line. The situation had been the same hundreds of years earlier under the Romans and throughout the Byzantine empire, and was similar to the situation that prevails today between Afghanistan and Pakistan. The treaty of Erzurum (1823) was also unsuccessful in clarifying the situation.

Such a situation was intolerable to both the Russians and British. Oil had not yet been discovered and they were not interested in the arid region itself, but they feared being drawn into a Turko–Persian local war when all their efforts were concentrated on consolidating their rule in the Caucasus and in India respectively. Accordingly, in 1843 the Russian and British authorities offered their services as mediators. With their help a more precise boundary was defined in a further treaty in 1847 and an Anglo–Russian boundary commission was appointed to survey the whole region in detail.

The wording of the 1847 treaty was still ambiguous and the commission worked slowly. They were not helped by an unfortunate incident in 1852 when all the British survey material was lost in the Thames off Gravesend, nor by the outbreak of hostilities between Russia and Great Britain in Crimea that occurred soon afterwards. By 1869, however, a detailed, though not particularly accurate, 60-foot long map of the border, formally entitled the *Carte Identique*, had been produced. This part of the map shows the northernmost section of the boundary with Mount Ararat, a place that had first featured centuries earlier on medieval world maps such as the Psalter map (p.78), surmounted by Noah's Ark.

Since the Russians and British were mediators and not arbitrators, they had no power to enforce their decision and thus had to rely on persuasion. The care taken in creating this map should be seen as a means of coaxing the Persians into accepting the proposed frontier line. All such efforts, however, were to no avail. A final settlement was only reached by Anglo–Russian arbitration in October 1914, just after the outbreak of the First World War.

Schofield (2006), pp.27–39.

Igor Ivanovich Chirikov, *Carte topographique de la frontière turco-persane par des officiers russes de la direction des commissaries-médiateurs Tchinkow et Williams*, St Petersburg: *Corps topographique impérial*, 1869
Manuscript, 113 x 99 cm
British Library Maps S.T.T.

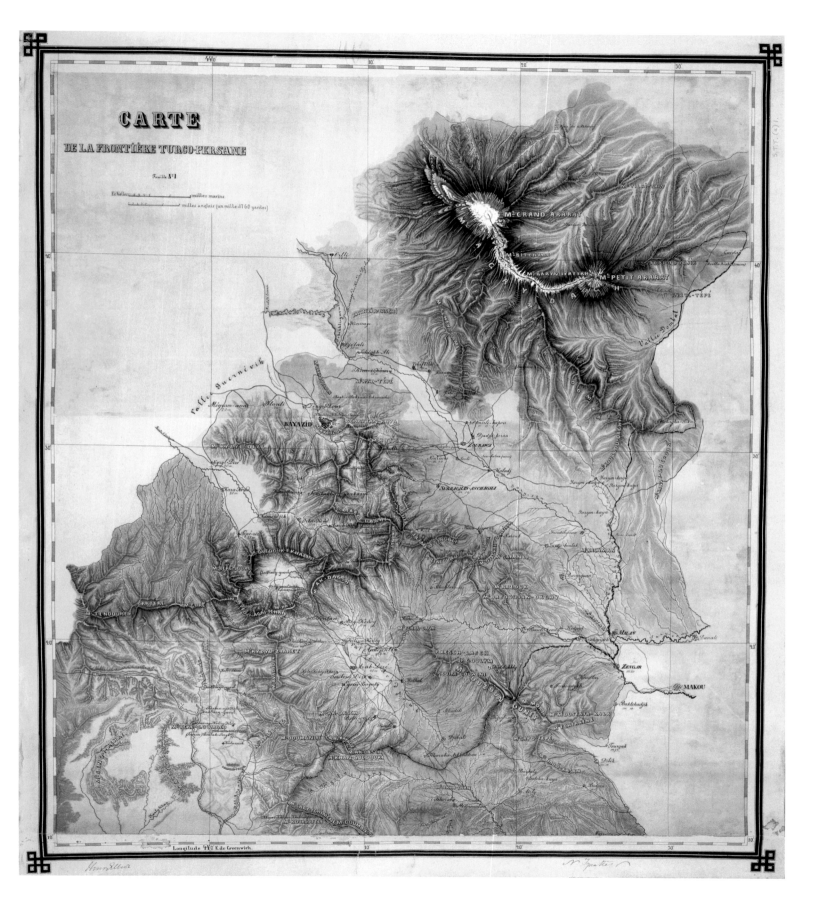

THE MERCHANT AND LANDOWNER'S HOUSE

The use of maps as wall decoration in private homes is probably best known through the paintings of Dutch seventeenth-century artists, notably Jan Vermeer, Willem Buytewech and Pieter de Hooch. These pictures need to be treated with a degree of caution, however. Contrary to the impression that they give, the relatively few inventories that have been studied for the maps mentioned in them suggest that most of the maps on display were not the large, multi-sheet maps seen in Vermeer's paintings. Much more common were modest, single-sheet, atlas-sized maps – of the sort usually displayed on the walls of map collectors' homes to this day – though they might have been enlarged through the addition of surrounding text. Some artists, for dramatic effect or in order to underline a wider philosophical or ideological message more effectively, even transformed what were actually folio-sized maps intended for atlases into large, multi-sheet maps in their paintings.[95]

Nevertheless, inventories make it clear that from the early sixteenth century large wall maps were frequently found in the homes of merchants and landowners (often the same people), even if an individual is unlikely to have had more than a few. From the late seventeenth century, newspaper and separately published advertisements give more detailed information about where in the houses the maps were likely to have been seen. In 1679, for instance, the London mapseller John Garrett advertised an oblong, four-sheet coloured wall map as being 'a fit ornament for a chimney piece', while another map in squarer format was suitable for a staircase landing.[96] A couple of decades later Philip Lea even suggested that maps printed on silk would make good blinds for sash windows, and some were doubtless used in this way.[97]

A few screens covered with maps and intended for use as draft excluders survive from the sixteenth, seventeenth and eighteenth centuries to demonstrate that the purchasers did take seriously the advertisements' advice that the maps could be used in this way. While the maps on most of these screens were individually selected by their owner and tended to be single-sheet examples made for atlases, some – for instance a series produced by the London mapseller George Willdey in about 1721, and repeatedly reprinted by later publishers – seem to have been created specifically so that they could be displayed on screens (pp.98–9).

The description of maps in inventories is frequently frustratingly vague, giving no indication of size or area depicted and almost never identifying the map's creator. A few conclusions can be drawn, however, from probate inventories taken in sixteenth-century Venice[98] and early seventeenth-century Antwerp,[99] as well as from stray comments in theoretical works on mapmaking and in contemporary diaries and memoirs, such as those of Samuel Pepys and John Evelyn. The first is that there was a difference between the display of maps in a country house and a town house.

In a country house, where there was usually more space, the display was more likely to resemble that in a palace, though with a lesser degree of luxury. Manuscript maps were likely to be restricted to estate plans, and unless the house was owned by a magnate or leading general or minister, none of the printed maps would be really enormous, though they might be assembled from several sheets. Such medium-sized printed wall maps can very occasionally still be seen in galleries of English stately homes, such as in The Vyne in Hampshire, where later editions of Ogilby's map of the City of London, originally published in 1676, and John Adams's route map of England and Wales of 1679 adorn the Stone Gallery. We also know that a suite of four tapestry maps – providing a panorama of England between Bristol and London, where the Sheldon family owned most of their land – used to adorn the principal room in Weston in Long Compton, Warwickshire (pp.56–9).[100] Unlike in a royal or municipal palace, however, maps were also likely to be found in the great hall, the first large room that a visitor passed through. They can still occasionally be seen there, as in Melford Hall in Suffolk. Here the hall is adorned with a large and early estate map of 1580 by Israel Amyce showing the lands of Sir William Cordell,[101] as well as another by Mark Pierce showing the same lands some fifty years later.

The entrance room – effectively the equivalent of a great hall or audience chamber in a palace – was also a frequent location for maps in town houses. Here maps of all sizes, and particularly world maps and maps of the continents would be displayed (pp.122–5). Other maps, including some quite small ones, were to be found scattered through the other rooms and corridors depending on the taste and interests of the owners. The parlour or study seems to have been a favourite location. While many maps were suspended on rollers, as shown in most Dutch paintings (which also reveal them, on closer examination, to be suffering the ravages of heat, light, dust and the effects of varnish),[102] many are described as being framed, and the frames seem often to have been made of quite luxurious materials, such as ebony.

There was a great variety in the size, quality and subject matter of the maps displayed in town houses. In the sixteenth century it seems that only ruling families and magnates could usually afford the largest printed maps, such as De' Barbari's famous map of Venice of 1500 (pp.52–3), regularly listed in the inventories of royal and princely palaces. Magnificent manuscript world maps seem also to have been restricted to the palaces of rulers and their most important advisers. Charles V, for example, presented large manuscript maps

Pieter de Hooch, *A Woman Drinking with Two Men*, 1658. National Portrait Gallery, London

OPPOSITE: Samuel van Hoogstraten, *View of a Corridor,* 1663. Dyrham Park, Gloucestershire

Jan Miense Molenaer, *The Artist's Studio*, 1631. Gemäldegalerie, Berlin

of the world not only to several Italian and German princes in the course of the 1520s, but also to cardinals and ministers, such as Baldassare Castiglione, diplomat and author of the influential tract *Il Cortegiano*.[103]

Merchants seem to have settled for more modest multi-sheet maps derived from the greater ones. Often they were woodcut rather than the more expensive copperplate. The first great map of London, the fifteen-sheet, so-called 'Copperplate Map' printed in about 1558, seems originally to have been intended only for presentation to royalty, courtiers and ministers able to aid the Hansa merchants, who seem to have commissioned and sponsored the map.[104] Merchants made do, as inventories show, with a reduced-size woodcut version, the so-called 'Agas' map. In early sixteenth-century Venice, the mapmaker Giovanni Andrea Vavassore seems to have earned his living by creating and selling such derivative, smaller maps, as to some extent did the Amsterdam map publishers Covens and Mortier in the eighteenth century.[105] However, by the seventeenth century, and still more so in the eighteenth century, increasing levels of prosperity throughout western Europe seem to have put very large, multi-sheet copperplate maps within the reach of wealthy merchants and landowners.

Judging from the inventories from Antwerp, the maps displayed in private houses in Catholic Flanders seem predominantly to have been of the owner's own country and its regions, and of his home town. Almost as frequent were world maps. Also common were maps of the Holy Land and of Jerusalem, while those of other countries and continents – with which the owner's ships presumably traded – were also to be found. Very occasionally there is a mention of a commemorative battle map, such as one of the siege of La Rochelle in 1629, perhaps by Callot, which in 1638 adorned the entrance hall of Andries de Laralde's house in Antwerp. Another, probably Callot's map of the siege of Breda, decorated the same space in Hendrik van der Goes's house in the same city in 1640 (pp.68–9).[106] As well as their decorative function such maps, commemorating Catholic victories over Protestant forces, also advertised their owner's political loyalties.

The Italian inventories and other sources examined by David Woodward suggest that the Italian merchant and landowner was far more cosmopolitan than his Flemish counterpart, though he chose the same rooms – notably the entrance hall and study – to display his best maps. As with the Flemish, world maps and maps of the four continents were very popular. Unlike the Antwerp merchants, however, the Venetians generally seem not to have displayed maps of their own city on their walls, leaving these to foreign visitors. Instead they showed more maps of other parts of the world, and of far-flung parts of their ever-diminishing maritime empire.

Rembrandt van Rijn, *Self-portrait*, 1661.
Kenwood House, London

coat of arms and to the selection and balance of colours used for the fields. Such maps successfully conveyed an image of prosperity achieved under the benevolent surveillance of the landlord, and they merited their place on the walls of his house – often far distant from the estates thus portrayed.

At the same time as the neat copy was delivered to the landowner, the more roughly executed final drafts of estate maps tended to be sent to the steward's office for everyday administrative use. A steward's room can still be seen at Felbrigg Hall in Norfolk where the large map, normally kept in its case high on the wall waiting to be pulled down, is accompanied by a way-wiser (a wheeled instrument used to measure distance), a surveyor's chain and other surveying instruments. Once the more elegantly executed estate maps had got dirty or damaged, or had been replaced by more recent surveys, they were often moved to the steward's office for practical use – where they would be annotated roughly and even cut up into sheets for convenience or to be given to the new owners when an estate was split up.[113]

If there was a manuscript map in an English merchant or landowner's town house, it was likely to be an estate map.[107] In 1653 the surveyor and theorist William Leybourne pointed out in *The Compleat Surveyor* that estate maps 'being well performed … will be a near Ornament for the Lord of the Manor to hang in his study, or other private place, so that at pleasure he may see his land before him'.[108] From the time that they first appeared in the mid-1570s, measured estate plans became ever more popular with the wealthiest landowners. Surveyors were soon complaining about the gigantic, multi-sheet maps that they were being commissioned to create. A few such monsters, for instance Agas's twenty-sheet survey of Lord Cheney's estates in Toddington in the British Library[109] still survive. Their intricate detail combined with their size make it doubtful that these maps were initially ever consulted for administrative purposes – they were more likely to have been admired as evidence of wealth and power. By the eighteenth century estate maps had become much less expensive and thus more common – and they sometimes depicted no more than the large back gardens of suburban villas.

Many plans were very plain – the hallmark of the estate maps by Christopher Saxton, creator of the first detailed printed English county maps, is their austerity.[110] Often the only decoration was to be found in the cartouches surrounding the title or scale-bar. Other maps might be gaudy or clumsily executed – not to be wondered at since surveying and drawing was only a part-time occupation for the almost all early local mapmakers.[111] A significant number of surveyors, however, possessed genuine artistic talent and considerable learning, and in the decoration of their estate maps they saw a chance to display both.[112] This was much to the liking of their patrons. The style might reveal its provincialism by lagging twenty or more years behind that of the artistic vanguard in London, but many maps were skilfully drawn and exquisitely coloured, with particular attention being paid to the owner's

95   Günter Schilder, *Monumenta Cartographica Neerlandica* vi (Aalphen-aan-den-Rijn, 2000), pp.47–8. Rembrandt's famous late self-portrait, now at Kenwood House in London, takes the process of cartographic manipulation for symbolic purposes still further. He pictures himself in front of a huge double hemisphere world map, stripped of everything except for the two circles. The composition recalls a detail from Jan Miense Molenaer's painting, *The Artist's Studio* (Gemäldegalerie, Berlin) showing an old man in front of a double hemisphere world map, in this case with the cartographic detail depicted.

96   R.A. Skelton, *County Atlases of the British Isles 1579–1850. A Bibliography 1579–1703* (Dover, Dawson, 1978), ill.27b.

97   Peter Barber, 'Necessary and Ornamental: Map Use in England under the Later Stuarts 1660–1714', *Eighteenth Century Life* 14/3 (1990), p2. And see Sarah Tyacke, *London Map-Sellers: A collection of advertisements for maps placed in the* London Gazette *1668–1719* (Tring, Map Collector Publications, 1978).

98   David Woodward, *Maps as Prints in the Italian Renaissance: Makers, Distributors and Consumers* (London, British Library, 1996), pp.79–93.

99   Schilder, *Monumenta Cartographica Neerlandica* vi, pp.50–4.

100 Hilary L. Turner, '"A wittie devise": the Sheldon tapestry maps belonging to the Bodleian Library, Oxford', *Bodleian Library Record* 17, no.5 (April 2002), pp.293–313; Hilary L. Turner 'The Sheldon Tapestry Maps: their Content and Context', *The Cartographic Journal* 40/1 (June 2003), pp.39–49; Hilary L. Turner, 'Tapestries once at Chastleton House and Their Influence on the Image of the Tapestries Called Sheldon: A Reassessment', *Antiquaries Journal* 88 (2008), 313–46.

101 A. Sarah Bendall, 'Pride of Ownership' in Peter Barber and Christopher Board (eds), *Tales from the Map Room*, (London, BBC, 1994), p.94.

102 Peter Barber, 'Profitable and Useful, Ornamental and Allegorical', in P. Barber (ed.), *The Map Book*, (London, Weidenfeld & Nicolson, 2005), pp.160–1.

103 Jerry Brotton, *Trading Territories: Mapping the Early Modern World* (London, Reaktion, 1997), pp.138–50. The magnificent map of 1525 by Diogo Ribeiro was sold just a few years ago by his descendants and is now in the Biblioteca Estense in Modena. And see Ernesto Milano, commentary to *Planisfero Castiglioni: carta del navegare universalissima et diligentissima* (Modena, Bulino, 2002).

104 Peter Barber, 'The Copperplate Map in Context' in Ann Saunders and John Schofield (eds), *Tudor London: a map and a view* (London, London Topographical Society, 2001), pp.16–32.

105 Van Egmond, passim.

106 Schilder, *Monumenta Cartographica Neerlandica* vi, p.53.

107 A sailor or naval administrator such as Samuel Pepys, or a council member of one of the great trading companies, might also have had some decorative manuscript sea charts on display.

108 For the development of estate plans see A. Sarah Bendall, *Maps, Land and Society: A History, with a carto-bibliography of Cambridgeshire Estate Plans c.1600–1836* (Cambridge, Cambridge University Press, 1992); David Fletcher, *The Emergence of Estate Maps: Christchurch Oxford, 1600–1840* (Oxford, Oxford University Press, 1995); Peter Barber, ' Mapmaking in England' in *The History of Cartography* iii, pp.1637–48.

109 British Library Add. MS 38065.

110 Several are illustrated in Ifor M. Evans and Heather Lawrence, *Christopher Saxton: Elizabethan Map-maker* (Wakefield, Wakefield Historical Publications and the Holland Press, 1979).

111 See the analytical introductory essay by A. Sarah Bendall (ed.) in *A Dictionary of Land Surveyors and Local Mapmakers of Great Britain and Ireland 1530–1850*, 2nd edition (London, British Library, 1997), i, pp.1–72.

112 Good examples of the range of artistic finish can be found in Margery Rose and Mary Ravenhill, *Devon maps and map-makers: manuscript maps before 1840* (Exeter, Devon and Cornwall Records Society, 2002).

113 See, for example, the maps of the Spencer estates throughout England in the British Library. Many of these maps suffered such indignities before being cut up and placed in huge make-volumes between corduroy covers bearing the family's arms (British Library, Add. MSS 78108–78155).

The world and four continents on a wall

The printed wall map became the specialist product of the cartographic publishing houses of Amsterdam. Such maps were expensive for the reason that, being larger than a regularly sized copperplate or a single sheet of paper, each had to be assembled from a number of smaller 'plates' and 'sheets'. These sheets, each bearing the impression of a portion of the map, would have then been assembled and stuck on canvas or cloth, coloured and varnished for display. Occasionally multi-sheet maps remained unassembled so that their detail could be conveniently studied. They were either stored in portfolios as separate sheets or bound up into books according to the requirements of the purchaser.

In their fully-assembled state, large wall maps appealed to the wealthy European mercantile class as a reflection of their social status and worldly knowledge. A set of five wall maps comprising the world and each of the four continents was a particularly popular product which each Amsterdam publisher would have displayed prominently in the window of his shop as the finest example of his skill. Not only would five large maps have covered a considerable amount of wall space, demonstrating the owner's possession of a sufficiently spacious house to accommodate them, but their scale would also have allowed for the most legible deployment of decorative and cartographic features.

The evidence of paintings of Dutch interiors indicate that it is unlikely a complete set of maps would actually have been displayed together in one room, even though their similar style and symbolism were designed to complement each other in a way not unlike a painted gallery. A set of six-sheet wall maps by Gerard Valk (1652–1726) of around 1680 provide a good example of this sort of intended cross-referencing.

The maps of Africa, America and Asia contain as part of their decorative scheme the portrait of the European monarch whose nation had been both historically and most recently involved there: a youthful Carlos II of Spain (1661–1700) in the Americas; Louis XIV of France (1638–1715) in Asia; and Charles II of England (1630–85) in Africa. To avoid any embarrassing differences in opinion, especially among Valk's prospective patrons, the continent of Europe remains unclaimed. The maps' Euro-centric attitude is perfectly understandable given the dynastic rivalries of the ruling houses and the almost exclusively European audience who would have owned the maps. The resulting vision is, of course, of a world controlled by Europe, its monarchies and governments, armies, culture and trade.

Printed wall maps had been produced in Germany, Italy and France from the early sixteenth century, and in England from the early seventeenth. But it was the Dutch wall map that came to be valued for its skilful manipulation of allegorical imagery and emotional baroque style, as well as its cartographic originality and superior technical quality. The Dutch had cornered the market by producing the quintessential and tailored cartographic wall display. Gerard Valk, one of the most active publishers in Amsterdam from the 1680s, formed a number of partnerships with other mapmakers such as Petrus Schenk (1661–1711), producing globes and updating and re-issuing older maps for publication. Valk's business continued in his family after his death, indicating the measure of success that he had enjoyed.

Shirley (2001), p.529 (no.532); Burden (2007), p.371 (no.675).

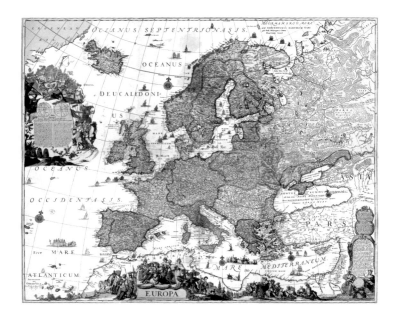

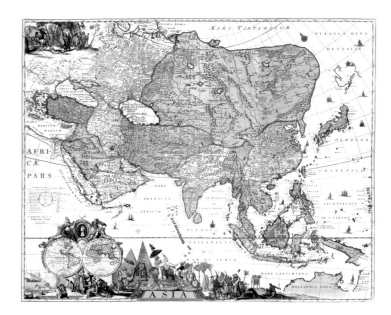

Gerard Valk, *Nova Totius Terrarum Orbis Tabula / Europa / Asia / Africa / America*, Amsterdam, *c*.1680

Five copperplate engravings each on six sheets, 118 x 135 cm

British Library Maps C.5.a.1

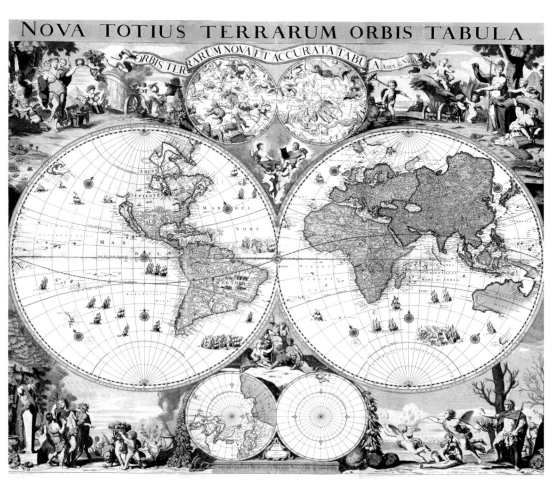

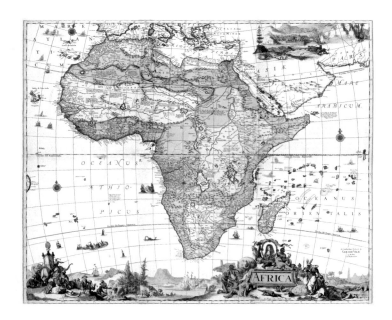

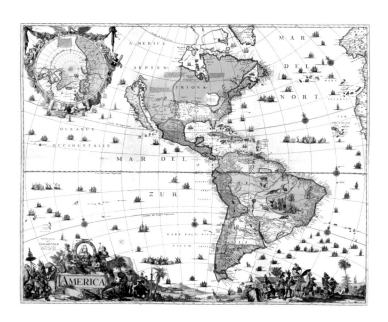

The currency of culture

The decorative double-hemisphere world map is occasionally seen decorating the walls of Dutch interiors in paintings of the seventeenth and eighteenth centuries. Although a great many smaller examples of the double-hemisphere map published in atlases survive, the larger, multi-sheet versions are less numerous. Examples in paintings by Jan Miense Molenaer and Rembrandt van Rijn are insufficiently detailed to be identifiable. Their evident popularity stemmed to a great extent from the complete, concise and, most importantly, attractive visual synopsis of the world and everything in it that they offered. By appropriating the space surrounding the hemispheres for inventive representations of parts of the world or mythological imagery often derived from paintings, the artist engraver was able to display his skill and appeal to the intellect of the educated viewer, who would have reflected on the symbolism and significance in much the same way as he or she would allegorical paintings. The map is also one of the great Dutch exports, with imitations appearing concurrently in England and France, and Japanese versions noted as late as the nineteenth century.

The Armenian world map by Adrian and Peter Schoonebeck, printed in Amsterdam in 1695, is of special significance. Not only does it inform us of the special importance of the Amsterdam map trade and the desirability of the Dutch 'product', it also illustrates the spread and prosperity of Armenian colonies in Western Europe. Double hemisphere wall maps were certainly not cheap, and to commission an eight-sheet map from one of Amsterdam's most renowned engravers would have been prohibitively expensive for all but the wealthiest of patrons.

The Armenian Archbishop T'ovmas Vanandetsi had set up an Armenian printing press in Amsterdam in 1688 with the purpose of producing literature to be sent to Armenians in Europe, Persia and Armenia itself. By the seventeenth century a network of Armenian colonies, founded originally as trading outposts, stretched from as far as China to Amsterdam. An Armenian hospice was founded in Bruges in 1478, and there are records of Armenian firms shipping goods on Dutch vessels in the early seventeenth century. Moreover, the role of Armenian businessmen as intermediaries between the trade of Europe and Asia had placed them in an influential position during the heyday of the Dutch East India Company, the headquarters of which were in Amsterdam.

Instead of adapting an existing Dutch map, a completely new map was commissioned because every word, name and letter had to be written in Armenian script if it was to be understood by an audience for whom neither Latin nor Dutch was their first language. Moreover the patron appears to have wanted the map to be particularly well understood, as each of the mythological figures is accompanied by a title. The figures, and the significance of their inclusion in the map, would have been readily appreciated by a West European audience, but not necessarily by those unfamiliar with the meanings of Western world maps. The purpose of the new map would appear to have been not only a visual cartographic object for a specialized market, but also an attempt to educate Armenians on the mannerisms of the predominant European culture. This aim was achieved, as copies of the map are recorded today in libraries as far afield as France, Armenia and Iran.

Shirley (2001), p.569 (no.575); Koeman (1967), pp.313–15; Papazian (1979).

Adrian and Peter Schoonebeck,
Double Hemisphere World Map in Armenian, Amsterdam, 1695
Copperplate engraving on eight sheets, 126 x 158 cm
British Library Maps *920.(89.)

The pious businessman's second home

The printed wall map of the Holy Land is documented in the family homes of seventeenth-century businessmen. It enjoyed a special place there, for just as a map of one's homeland might symbolize pride, or a map of a distant region might represent the source of one's wealth, the map of Palestine communicated one's piety, knowledge and religious orthodoxy.

To understand the popularity of Holy Land wall maps, it is necessary to appreciate the nature of religious observance in Renaissance Europe. The Protestant Reformation, which began in Saxony in 1517, had elicited a stricter emphasis upon personal study of The Bible, and maps of the Holy Land came to assume a particular significance as didactic tools. One of the major features of these maps was the incorporation of Biblical events and narratives, such as the Exodus and the Nativity, shown in their correct geographical locations along with key quotations from scripture. As well as providing an educational opportunity for children within the home, the map would have served to dispel any dangerous accusations of impiety on the part of the owner. Later such maps functioned in the same way for Counter-Reformation Catholics. The genesis of this map of 1570 was in the European empire of Catholic Spain, having been produced by Christian Sgrooten (1532–1608), cartographer to Philip II, and published in Antwerp by Hieronymus Cock (1507/10–1570).

The sheer number of wall maps of the Holy Land which survive, or are known through documentary sources, confirms their popularity over all but perhaps world maps. The scarcity of individual surviving examples, however, suggests that most of them would have been placed on walls and soon been destroyed, for it would clearly not be economically viable for publishers to print only a small number of copies. The Holy Land map by Lucas Cranach the Elder of 1515, for example, survives in a single example, as does Gerardus Mercator's large map of 1537, while a unique sheet of one 12-sheet Dutch map of the Holy Land from 1538 was recently identified in the British Library. Sgrooten's map of 1570, printed on nine sheets, is also a unique survivor. Sgrooten based his map on the accounts of Peter Laicksteen (fl.c.1556–1570), who had made a pilgrimage to the Holy Land in 1556. Such pilgrimage accounts tended to be more concerned with the correct location of religious episodes than describing the terrain, and it is for this reason that Holy Land maps were able to function as unimpeachable reflections of their owners piety.

Karrow (1992), pp.329–31; Nebenzahl (1986), pp.82–3 (no.20); Delano-Smith and Morley-Ingram (1991).

Christian Sgrooten / Peter Laicksteen / Hieronymus Cock,
Nova Descriptio Amplissimae Sanctae Terrae, Antwerp, 1570
Copperplate engraving on nine sheets, 103 x 108 cm
British Library Maps C.10.b.2

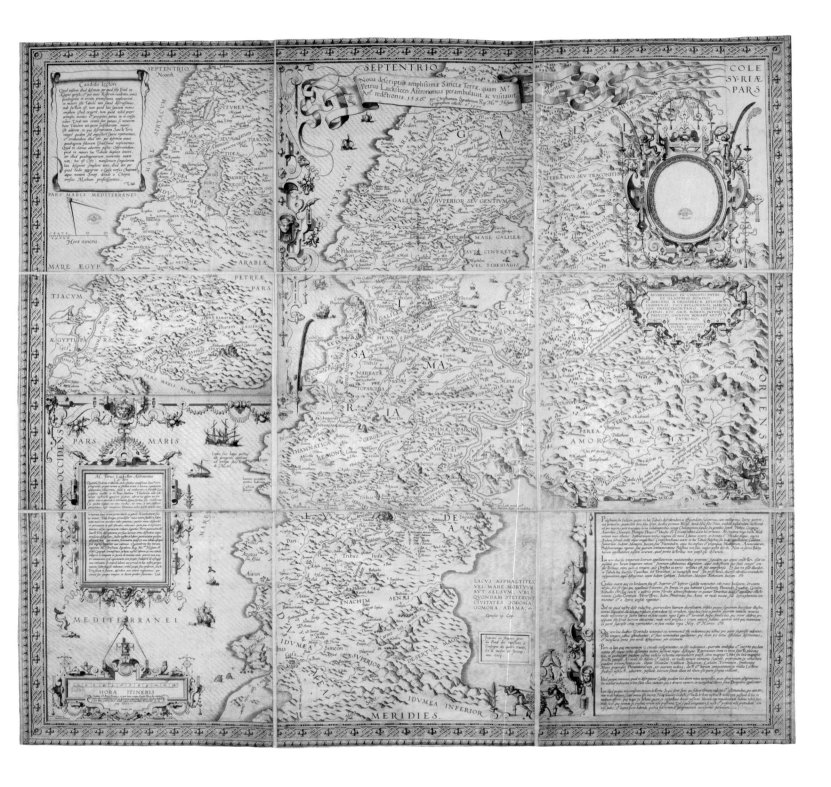

The European world view

Although printed wall maps of the four continents were produced as sets by seventeenth-century map publishing houses, their size usually prevented them from being displayed together in the modestly-sized town houses of most merchants. Instead they tended to be hung in different rooms, according to the tastes and priorities of the owner. Of the four continents, it is unsurprising that the map of Europe would have had special significance in the parlours of Rotterdam and Seville, perhaps enjoying the central location above the fireplace where, through extremities of temperature and soot, in addition to varnishing and exposure to sunlight, it stood little chance of survival. The British Library's Europe map of 1617 by Johannes Janssonius (1588–1664) is the only surviving copy. It is unusual in that it has not been coloured, backed with canvas or mounted on a wall, yet this has allowed a clearer view of the sophisticated language which was designed to enlighten and flatter its affluent owner.

The symbolism of Janssonius's map functions by appealing to a continental as opposed to a national or civic loyalty. Though the map is dedicated to Louis XIII of France, the dedication is subdued. The vision it imparts is of a Europe divorced from common national and dynastic rivalries, and with an ancient cultural and religious heritage, rich natural beauty, abundance and wealth, and abounding in art and science. The perceived pre-eminence of Europe, described in the map as 'the Mother and Queen of the whole world', was as much a way of flattering the prospective European owner as a celebration of the continent. Furthermore, the map imparts the idea of a shared European identity that would have been appreciated by tradesmen from rival European nations, who nevertheless viewed each other as allies when conducting business far from home. While the racist connotations of the map are implicit rather than direct, they would have been favourably received by the owner in the same way as those contained in maps of Africa, America and Asia which similarly expounded the European world-view of superiority and conquest.

A number of European trading companies were international, yet the business of the majority of merchants, and especially those of the Hanseatic League and the Merchant Adventurers, was primarily European in scope. By the seventeenth century their trade was focusing on the newly opened-up waters of the Baltic Sea and the vast new markets of Scandinavia, the Baltic States and Russia. In this context it is significant that the peculiar elongated projection on which the map is drawn serves to accentuate the northern waters of Europe. To a trader of the Hanseatic League, and many a Dutch merchant, the Baltic Sea would have been recognized as rivalling in size, and thus in importance, the Mediterranean Sea.

This map published by Janssonius not only reflected the owner's business interests and his sense of pride: it also appealed to his intellect. The inclusion of such elements as the projection and rhumb lines (a navigation-assisting network of lines drawn from points on the compass) would have satisfied his scientific self in the same way that the complex (and incomplete) group of mystical and seemingly unrelated scenes along the top of the map would have piqued his curiosity.

Schilder (2003), pp.337–45.

Cornelis Claesz /Johannes Janssonius, *Europa*, Amsterdam, 1617
Copperplate engraving on eight sheets, 94 x 145 cm
British Library Maps S.T.E.

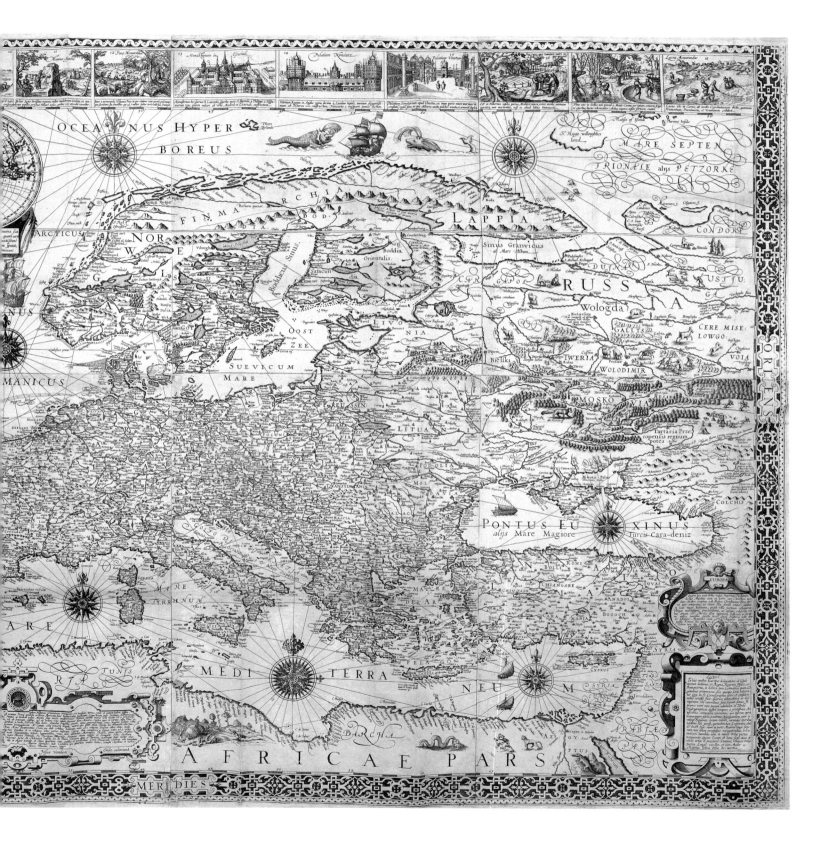

A step on the ladder

A wall map depicting a single English county was the ultimate cartographic expression of eighteenth-century provincial pride. Combining the elements conducive to good mapmaking – accurate measurement, large size, much detail, skilled engraving and artistry with reference to the history, heritage and people of the county – such a map served as a status symbol, putting other counties in the shade. The map was an emblem of community, wealth and trade. It could also be a measure of an individual's success.

Large-scale county maps were almost prohibitively expensive to produce and to purchase. The standard method of financing them had long been by means of subscription, whereby the buyer paid half of the total price in advance and half upon receipt of the map. As an additional incentive, the publisher of this map of Sussex invited potential buyers 'who are to have their arms engraved therein to send them to Bernard Lintott, bookseller, and … the said map shall be immediately published'.

The 184 arms adorning the map constitute a veritable who's who of the county of Sussex in the early 1720s. The arms of the dukes, earls, viscounts and bishops are arranged according to the age and importance of their titles. Below these are listed, in alphabetical order, everybody who had submitted their arms to the publisher in time. Included are esquires, gentlemen, baronets, reverends, rectors, sires, captains and at least one major-general: pillars of communities and, in some cases, of national standing. The arms of Henry Pelham (1694–1754), for example, at the time MP for Sussex and secretary at war (and to become prime minister of Great Britain from 1743), sits among the arms of a further four Pelham family branches. Next to the illustrious Pelham group is a Mr Samuel Plummer, gentleman, from the village of Beech.

An apparently fixed hierarchy of status was not an insurmountable obstacle to the successful, upwardly mobile businessman. Henry Lintott, whose arms are out of sequence (suggesting that it was one of those added later to a vacant space), was the brother of the printer and came from an artisan background, but in 1736 he would be promoted to the position of sheriff of the county of Sussex. Aspiring businessmen were able to rise up the social ladder, and the printed map helped to underpin their success by impressing those who saw their name and arms, and the location of their homes (keyed by co-ordinates to the appropriate place on the map), rubbing shoulders with the great of the county.

The map by Richard Budgen (1695–1731) is one of the earliest large-scale county maps of England. It was produced in the wake of many derivative atlas maps of the preceding century, but anticipates the large-scale mapping encouraged by the financial rewards offered by the Society of Arts from 1769. The map integrates enough fresh information for it to be considered independent from earlier maps, and it compares favourably in terms of content and accuracy with the map of Sussex by Thomas Yeakell and William Gardner in 1778–83. Budgen was from a family of estate surveyors and local mapmakers; his grandson Thomas would become one of the first surveyors of the Board of Ordnance some seventy years later.

This copy of Budgen's map, an example of the second state (containing amendments not present in its previous printing) dated 1724, comes from the geographical collections of King George III. The thick and liberally applied paint in a number of different hues reinforces the impression of its royal provenance.

Kingsley (1982), pp.57–63.

Richard Budgen, *An Actual Survey of the County of Sussex...*, London, 1724–5
Copperplate engraving on six sheets, 104 x 153 cm
British Library Maps K.Top 43.3.8 TAB END

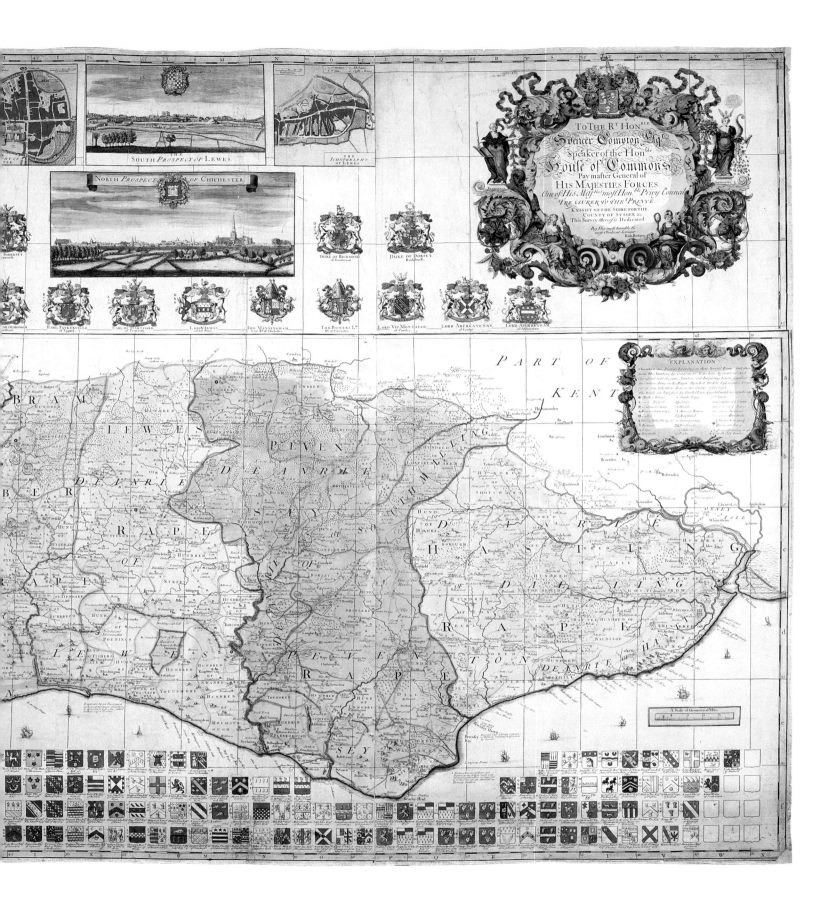

Royal London

This map of London can only fully be understood in the context of the political situation prevailing at the time. It was published in the early 1680s in the months after Charles II effectively staged a coup against a loose political grouping, the Whigs (in some senses the successors to the Puritans), who had been making the political running over the previous five years. The map commemorates and immortalises the royalist reaction.

The map itself is of exceptional technical importance. It was based on the first detailed and truly scientific surveys of the City, Westminster and Southwark, which had been underway since immediately after the Great Fire of 1666. The surveys had been undertaken by some of the country's leading scientists, supervised by Robert Hooke on behalf of the newly founded Royal Society. For the first time virtually the whole of the built-up area was depicted in plan, showing the layout of buildings vertically, and evidently on the basis of mathematical calculation, rather than pictorially and seemingly non-mathematically as a bird's-eye view, as had previously been popular. It was partly meant to demonstrate that the British could match the French achievements in scientifically based cartography encouraged under Louis XIV and his minister Colbert.

It is, however, the map's decoration that is striking to the eye. The handsome and very accurate panorama of the city as a whole is faultless, as is the deviation from pure science that led what were judged to be the most important buildings to be shown in elevation. This was to be expected in a map meant primarily for display. It is the illustrations, however, that reveal the political agenda. Nowhere had the Whigs, who tended not to conform to the Anglican church, been stronger than in London, and the decoration emphasizes their recent political humiliation. While images of such important mercantile buildings as the Royal Exchange, Guildhall (then as now the City's town hall) and Mercers' Hall are featured, the bulk of the illustrations show buildings relating to the monarchy and the Anglican establishment.

Charles II and his queen are prominently seen at the top left next to a view of the so-called 'Holbein' Gate in Whitehall Palace, receiving the subscription book of the survey from Morgan's deceased associate (and step-grandfather), the Scotsman John Ogilby. There are views of the palaces of Whitehall and Westminster, of the queen's palace of Somerset House, and of Westminster Abbey. A yet-to-be rebuilt St Paul's has been given a dome that was never built to this design. In the upper corners are equestrian statues of Charles II and his father, while the names of all those who had subsidized the survey are listed or seen hanging from grape vines or pear trees (perhaps alluding to the their role in nurturing the creation of the map). With princes, privy councillors, ministers, peers, bishops, masters of Cambridge and Oxford colleges and office-holders of all sorts, as well as the mayor and some city aldermen, they represent the ruling elite, emerging triumphant from years of Whig harassment.

The inscriptions on the map exhort the king to 'befriend your Allies & Punish your Enemies, succour the Distressed & Defend ye truly Antient Catholique Apostolique Faith. May the great God preserve & suddenly Convert or Confound all your Enemies'. No wonder that Charles commanded that this map be displayed in the halls of city livery companies to ram home the new reality of royal power. No wonder also that the more compliant city merchants displayed the map in their homes to demonstrate their wealth and pride in their city – and their political loyalty.

Hyde (1977).

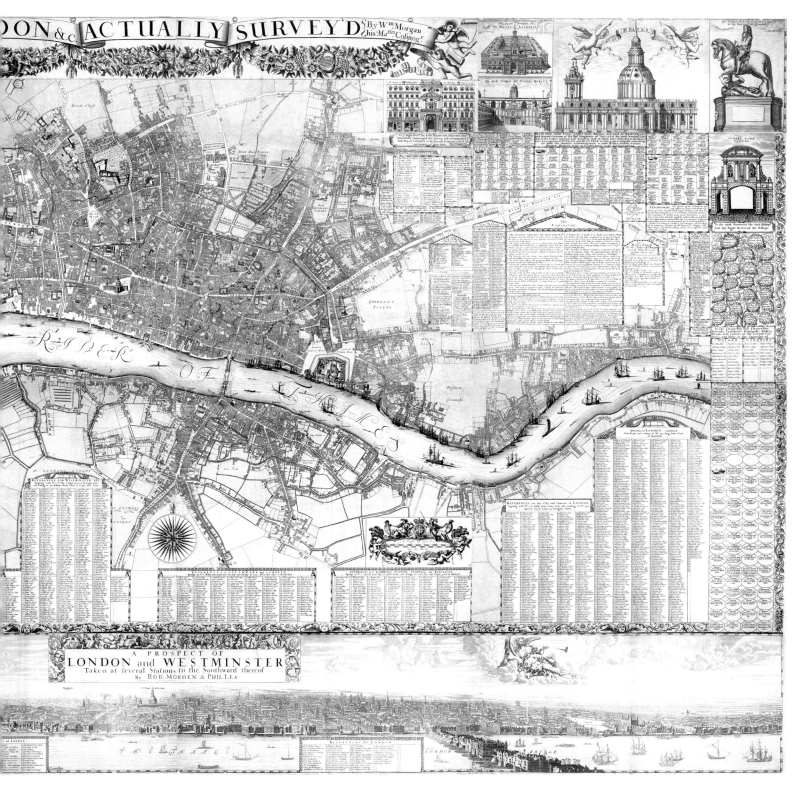

William Morgan, *London &c actually surveyed*, London, 1682

Copperplate engraving on twelve sheets, 160 x 250 cm

British Library Maps Crace Portf. II. 58

The insider's tourist map

At first glance one could be mistaken for thinking this a map of the Isle of Wight. In fact Stephen Walter's 2008 map of London, entitled *The Island*, provides an interesting take on the London-centric view of the English capital city as independent from the rest of the country, while alluding to the fact that the United Kingdom is itself made up of a collection of islands. Equally strong is the map's display of a level of local knowledge and pride – crucial both to the artist, a native Londoner with an 'ingrown passion for the city,' and to the viewer, who requires a level of knowledge to appreciate fully both the tone and the depth of information from which the map is constructed.

The map focuses upon Greater London and its outlying commuter towns in impressive detail. It presents a varied range of local and personal information in words and symbols, including the locations of pubs with good views, the ethnic make-up of areas, notable residents, speed limits, ice-cream vendors and the sites of famous (or infamous) events. In a sense, the map and the language it employs bear similarities with the modern tourist map, pointing out features of interest for the curious. Yet if the map can serve as a tourist map, then it is one for the insider: the Londoner who knows, or who has heard, where the best meals, the whereabouts of danger-ous dogs or the best venues for outdoor copulation are to be found. This is, of course, not information that would find its way onto any conventional map, and one cannot imagine such a map being officially commissioned – just as one cannot imagine William Morgan's 1682 map of London giving special mention to the most squalid areas around the Fleet prison.

Comparison of Walter's map with other earlier maps of London (a number of which the artist acknowledges here) reveals interesting points about the types of people who would have viewed them. The audience for Morgan's seventeenth-century map, for example, was primarily official or business based, and accordingly his map shows the city as prosperous, elegant and fully recovered from the fire of 1666. There are obvious similarities with Charles Booth's 1889 *Descriptive Map of London Poverty*, which represents wealth and class by area, but perhaps the closest map to Walter's *The Island* is MacDonald Gill's 1914 *Wonderground Map of London*, which is constructed of a number of pictorial puns, comments and slogans reflecting the significance of particular areas to the society of early twentieth-century London. The humour and symbolism of Gill's map is virtually incomprehensible to the modern viewer today, and the same may well be said of Walter's map a century from now. Nevertheless, both maps function by speaking to the initiated in the dialect of their times, focusing on what was thought of as important, interesting or, in Walter's case, shocking or mundane.

Whereas the *Wonderground Map* was produced as promotional material for the London Underground, *The Island* is a limited edition work sold on the art market to a limited clientele (unsurprisingly the map shows contemporary art spaces and galleries). This should not lead us to assume the exclusiveness of the map. One does not have to own something to understand or appreciate it, as those adopted Londoners who have spotted

Stephen Walter, *The Island*, London, 2008
Inkjet print, 140 x 200 cm
British Library Maps CC.6.a.30

134

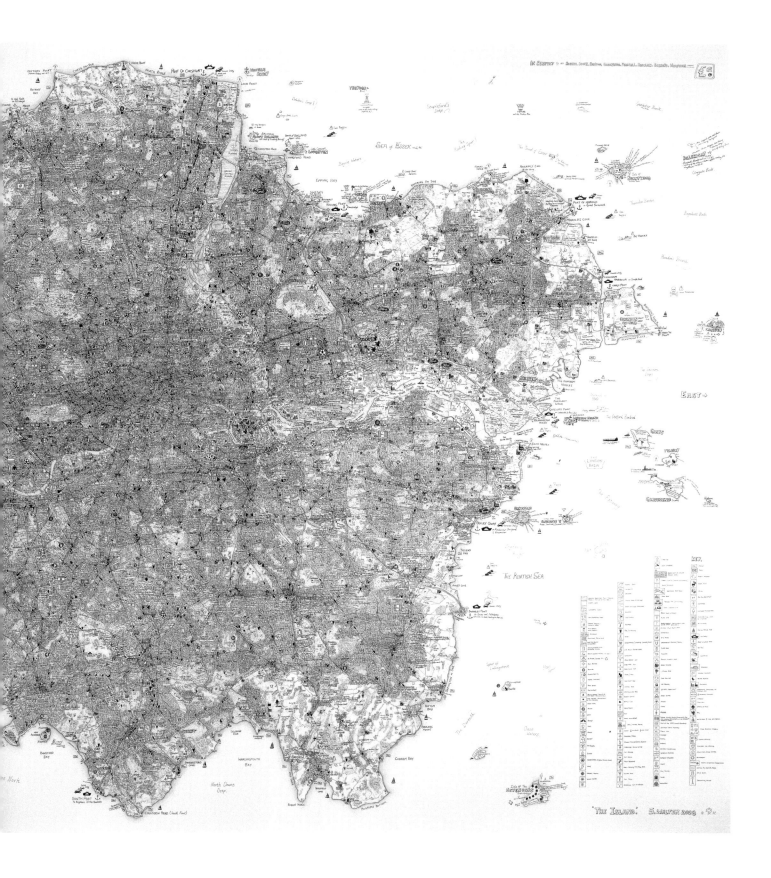

the artist's map of Islington hanging appropriately in a public house off Upper Street will testify.

Stephen Walter writes: 'After producing a map of the UK and Ireland I came to the decision that I would make one of London. With an ingrown passion for the city as a native Londoner, I began this undertaking in 2006. It was to span over two years.

'It was drawn to the limits of illegibility and required the use of a magnifying glass to produce and to read it (the detail was slightly englarged in the prints). Using a projection from a map of the city I began with a basic trace of the outlines, main roads and railways. The rest was done totally free-hand through the use of other existing maps. Once the basic geographical and historical information was laid down through a couple of informal top-left to bottom-right processes, I was able to float from one area of the map to another in order to fill in the detail, as and when it came to me. A year-and-a-half later, after much research and application, the drawing was complete. It took another six months to formulate the work into an exhibition of thirty-four prints, each one dissected from the original scan.'

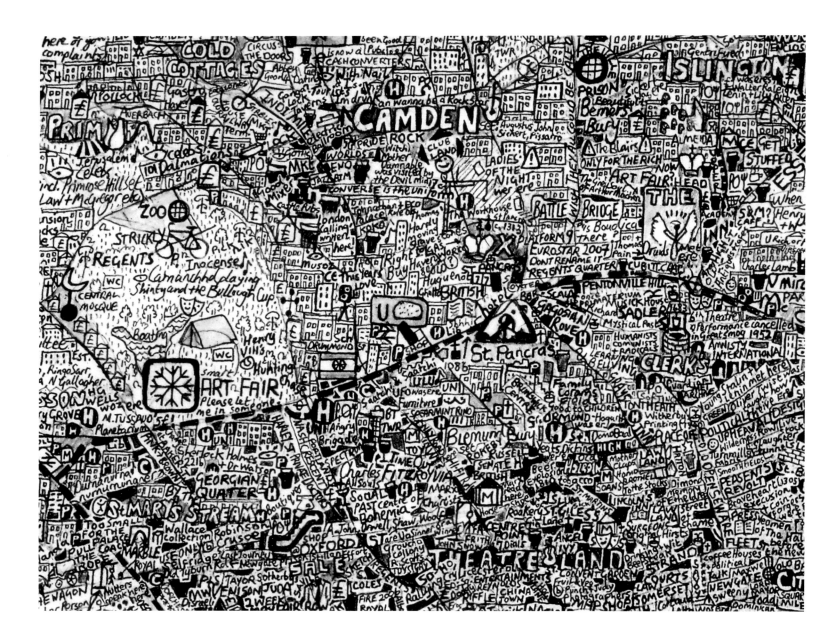

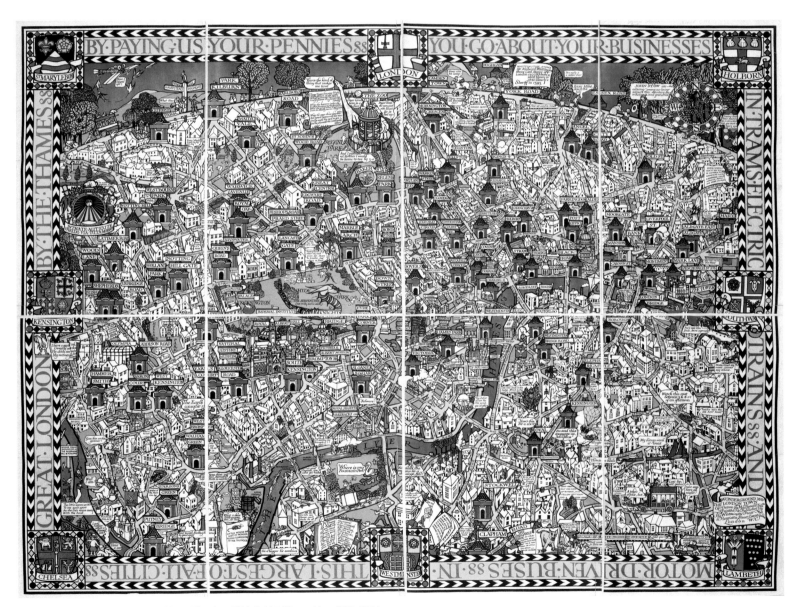

MacDonald Gill, *The Wonderground Map of London*, 1914. British Library Maps 3485.(199.)

The pride of ownership

This splendid map of an estate northeast of Norwich (which at that time lay on the edge of the fens) is drawn to a consistent scale of 1: 2376 (nearly 50 cm to the kilometre or about 26 inches to the mile). It was created by a cultured Suffolk gentleman, John Darby (d. 1609), only seven years after the earliest known example of an estate map drawn to a consistent scale, which also showed a Norfolk estate. Darby went on to create several other very decorative local maps. It may be significant that the patron, Sir Philip Parker, identifiable from his arms, seems to have known Thomas Seckford, the patron of Christopher Saxton – creator just three years earlier of the first printed atlas of English and Welsh county maps.

This map was expensive to produce (though we do not know its exact price), and it is far more handsome than was strictly necessary for land management. It was almost certainly created for display in Sir Philip's newly built residence at Erwarton, southeast of Ipswich in Suffolk. Even though he owned most of the manor of Smallburgh, Sir Philip had no home there and presumably wanted to admire – and impress his visitors with – the extent of this and all his other estates in the comfort of his main home.

The map is cartographically sophisticated and in some ways trail-blazing, dating as it does from a period before conventions for estate maps became formalized. Comparison with modern maps shows the survey was mathematically accurate. The mapmaker has used a range of attractive colours to distinguish systematically between various sorts of land-use: marshland, heath and meadow, pasture and arable. He captures well the mixture of enclosed and open-field arable (strip) farming, and meadowland and marsh that characterized Elizabethan Norfolk.

It is, however, the decoration that is most striking, depicting as it does a range of agricultural pursuits, from farmyard activities to hunting and shooting, and a variety of animals. Particularly noteworthy are the large figures at the lower left and right, likely to be copied from prints after Pieter Brueghel. The postures of the cherubs above the direction indicators also seem to be vaguely influenced by prints of Michelangelo's Medici Tombs in Florence. Portrayal of the strolling beggar at the bottom left, as well as the relative prominence of Smallburgh church, were probably politi-cally motivated. Most of Sir Philip's family was Catholic and he, a sheriff of Suffolk, needed to demonstrate his loyalty to the Anglican establishment. The beggar with a monkey on his shoulder also could be a satirical reference to the former head of the family, his half-brother, the recently-deceased Lord Morley – a penniless religious exile thought to have dabbled in treason while wandering fairly aimlessly for many years across southern Europe. The church served quietly to emphasize Sir Philip's personal Anglican orthodoxy.

The fresh quality of the map is due to the fact that it is unfinished and so was probably never displayed, although later annotations show that it was being used for land management purposes as late as 1762.

Barber (2005), pp.55–8; Bendall (1997), D.035.7.

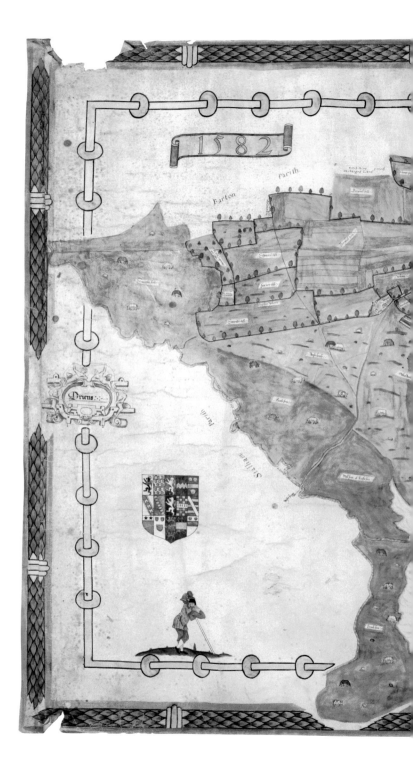

John Darby, *Plan of the Parish of Smallburgh*, Bramford, Suffolk, 1582
Manuscript on vellum, 104 x 177 cm
British Library Maps Roll 527
Purchased with assistance from The Friends of the British Library, 2004

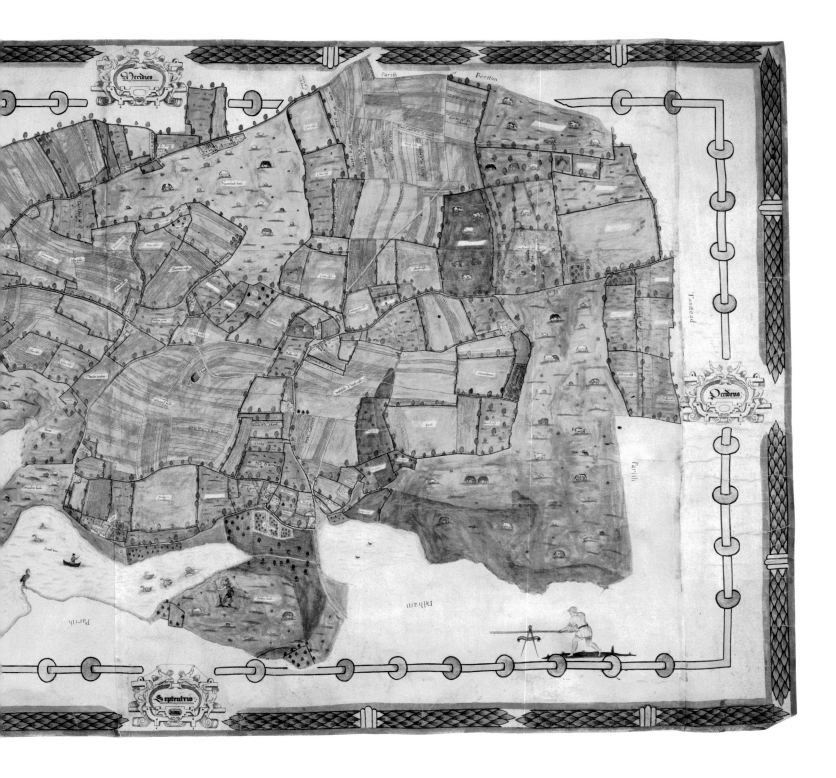

William Penn's neighbour

Francis Hill, *A Map And description of all ye Lands belonging to Richd Bridger Esqr in ye Parish of Worminghurst*, Canterbury, 1707

Manuscript on vellum, 66 x 167 cm

British Library Add. MS 37420

Richard Bridger's family had been lords of the manor of the little village of Warminghurst in West Sussex for seventy years by 1707, when he commissioned this map. Richard did not live in the manor, however. Its mansion house, Warmington Place, and surrounding land had been separated from the manor and sold off over a century earlier. Yet as the map shows, Bridger continued to own the bulk of the manorial land which was leased out.

The map was surveyed and presumably drawn by Francis Hill (d. 1711), a member of a family of land surveyors based in Canterbury. It names adjacent owners, uses astronomical signs to show which of the tenants occupied specific fields, indicates the site of the watermill and shows tracks of all sorts. In particular Hill distinguishes between bridleways where only horses could go ('wapple'-ways) and those accessible to carts as

well ('sheere'-ways). He was doubtless paid well to create a map which was far more opulent than the normal run of his work, while a utilitarian and less decorative draft was probably handed to Bridger's steward for practical use in running the estate. The elaborate map was probably intended to adorn the walls of Richard Bridger's main residence, perhaps with similar maps of his other lands, to impress visitors with the extent of his landed wealth.

The quality of execution is superb and the lavish decoration was carefully selected. Larger counterparts of the semi-nude figures lining the title cartouche would have been found, in stone or plaster, supporting the hall ceilings of the grandest and most fashionable baroque palaces. Tulips, placed in small blue-and-white porcelain vases, adorned the best rooms

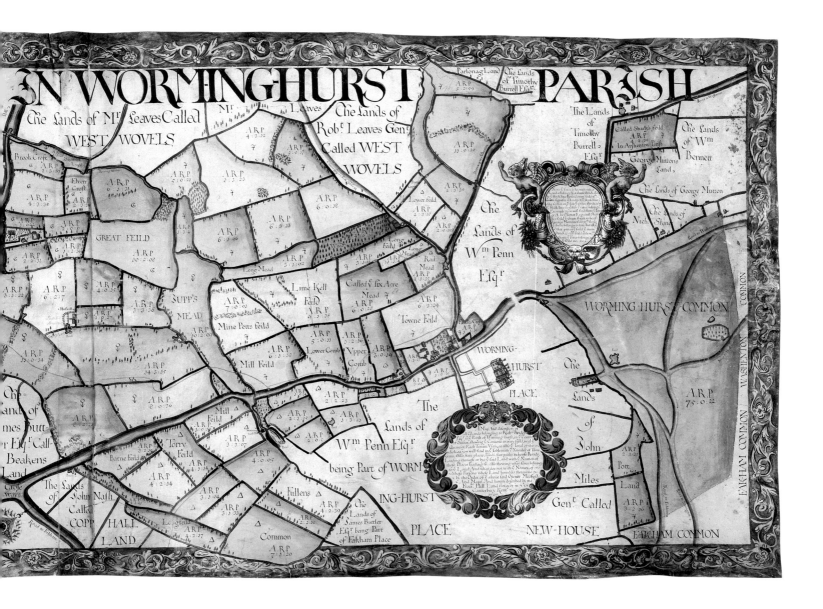

in homes and palaces, from Hampton Court and Kensington Palace downwards. The acutely observed goldfinch and greenfinch betrayed an equally fashionable, quasi-scientific, interest in the accurate observation of nature, while the two compasses carelessly strewn above the scale bar further reflect Bridger's interest – genuine or feigned – in science. The monogram discreetly placed between the birds is that of Queen Anne herself, a subtle reference to Bridger's loyalty to the Protestant Succession in Great Britain. Dominating the composition at the top left is his coat of arms, coupled with those of his wife Elizabeth, daughter of John Alford of Somerset.

Few buildings are shown, but among those featured are the church, a further reminder of Butler's loyalty to the established church and

Warminghurst Place, even though it was then owned by William Penn. It is this old house, to be demolished shortly afterwards, that is of most interest nowadays. This is the only known image of a building already ancient when it became home to Penn, from 1676, and it was here that he planned and worked out the constitution of his colony of Pennsylvania.

In 1707 Penn was forced to sell the house to raise money. This map may have been commissioned to clarify the boundaries between Penn and Bridger's lands in connection with the impending sale.

Steer (1961), p.26; *Victoria County History of England. Sussex* vi, (1986), pp.52–4; Bendall (1997), H.358.

Business concluded

Just as maps have played a part in business proposals and promotions, they have also fulfilled a retrospective commemorative function for those involved in bringing a successful business venture about.

The settling and colonization of 45,000 acres of land in North America known as Pennsylvania followed the grant of land in 1681 by Charles II of England to William Penn (1644–1718). By any standards, it is one of the largest and most significant business ventures in history. Penn, a Quaker, came from an English gentry family with land interests in Kent. He had managed to secure the grant through his influence and patronage, and he intended to resettle people of a similar Puritan persuasion in the face of discrimination in Europe. But although his colonizing mission was an evangelical one, it did not preclude making money.

The plan for populating Pennsylvania with European settlers involved maps along every step of the way: maps for planning purposes, maps for advertising and promotion, and maps for display by those who came to own land. A map of the projected town of Philadelphia, for example, designed by Thomas Holme (1624–95) and published in *Portraiture of Philadelphia* 1683, was used in promotional material in England, Holland and Germany. Holme was the first surveyor-general of Pennsylvania, working alongside Penn, and his maps are inseparable from Penn's role in overseeing the realization of his vision.

The *Map of the Settled Part of Pennsylvania* was not part of the speculative or planning process. Instead it was a commemorative work, intended to celebrate the settling of Pennsylvania and the success of Penn's project within the short space of six years. It contains the names of the 700 landowners, each placed over their particular plot of land; but significant to understanding the purpose of Holme's map is the fact that only around half of these landowners actually settled in Pennsylvania. For those who chose not to emigrate to America, the map would have constituted proof of their newly acquired land, to be displayed proudly on the walls of their homes in Britain or mainland Europe.

Holme's map was engraved by Francis Lamb and sold by John Thornton and Robert Greene in London. Although four examples survive, this is the only remaining copy of the map in its earliest form, published in 1687–8. It comes from the collection of Sir Hans Sloane (1660–1753), the effective founder of the British Museum and an acquaintance of William Penn with an interest in the natural history of North America. Although Sloane did not own any land in Pennsylvania, the borders of the map indicate that it was once attached to a wall.

Burden (2007), pp.299–301 (no.628); Snyder (1975).

Thomas Holme, *A Map of the Settled Part of Pennsylvania*, London, 1687–8
Copperplate engraving on seven sheets, 102 x 149 cm
British Library Add. MS 5414.23

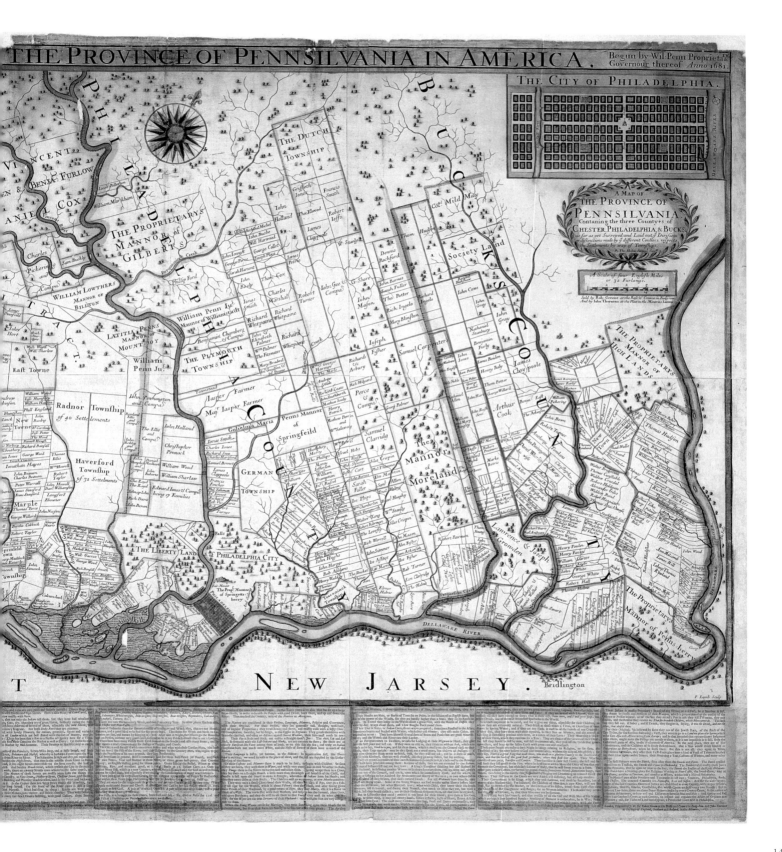

Where to send your fishing fleet

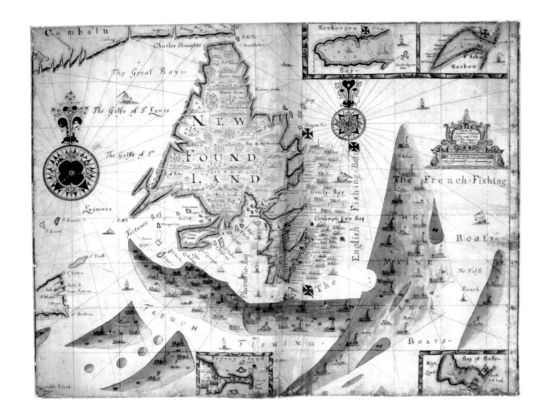

Augustine Fitzhugh, *Chart of Newfoundland and the Grand Banks*, 'next Doore to the Shipp in Virgine Street', Wapping, 1693
Manuscript, 101 x 128 cm
British Library Add. MS 5414.30

Until very recently fishing and Newfoundland were synonymous. The abundance of fish to be found off the Grand Banks probably drew Bristol mariners to Newfoundland before the island's official 'discovery' by John Cabot in 1497. Over the following centuries the area became a magnet for sailors from Spain, Portugal, England and France. By the 1690s Iberian fishing fleets had effectively been excluded, and there was fierce competition between English and French fleets for the fishing stocks. Salted fish, especially cod, from the Grand Banks provided a significant proportion of the English diet.

This map, from the collection of Sir Hans Sloane (1660–1753), who had a long-standing interest in North America and the Caribbean, would have provided essential information for any merchant investing in English fishing fleets. The fact that it is manuscript and not printed suggests that its circulation was meant to be restricted: 'Commercial – in confidence' we would label it today. The map's emphasis is on the Grand Banks, coloured in yellow, with those parts that were above the waterline and dry distinguished in a darker hue. The plentiful soundings were intended to prevent the ships from running aground, and the map helpfully points out the 'Falss Banck'. This copy, on paper rather than the tougher and more usual vellum, was probably meant for the merchant's chamber rather than for use on board ship.

The map depicts 'English Fishing Boats' as being confined to a relatively small area off the east coast of Newfoundland, while 'French Fishing Boats' enjoy the best of the fishing to the south and east. The map was created when England was at war with France, and it might have been intended to persuade government ministers that this unsatisfactory situation should be reversed – as it eventually was, under the terms of the treaties of Utrecht (1713) and Paris (1763). Newfoundland was at that time largely uninhabited apart from a few forts and small settlements intended to service the fishing industry.

The chart was created in Wapping 'next Doore to the Shipp in Virgine [i.e. Virginia] Street', which still survives. Its maker, Augustine Fitzhugh, belonged to a school of chartmakers, most of them members of the Drapers Company of London, which had been working in Docklands (near the shores of the Thames, east of the Tower of London) since the 1590s. Fitzhugh had been apprenticed in 1673 to John Thornton. Fitzhugh's style is typical of the 'Thames School', with elaborate windroses often (as here) incorporating an English rose, its touches of gold, and accurate nautical information concealed by the rather naive renderings of ships, flags and forts. The scale bar is, appropriately enough, flanked by two garishly coloured fish.

Campbell (1973); Smith (1978), pp.45–100.

For the foreign market

A view of Guangzhou (Canton), representing the central portion of the city, Canton, *c.*1770
Manuscript, 100 x 180 cm
British Library Maps
K.Top 116.22.2 TAB

This bird's-eye view of Guangzhou (Canton) by an anonymous Chinese artist is an export painting, produced exclusively for the European merchant wishing to display a memento of his travels and represent the basis of his wealth. Paradoxically, although the novelty of a view drawn in an eastern manner would have alluded to the owner's appreciation of different cultures, it was fashioned expressly to resemble a European creation. The resulting image is an intriguing mixture of artistic styles.

European trade with Canton had begun with the English East India Company in around 1700, and by the end of the eighteenth century the Western presence there comprised Dutch, French, Spanish, Danish, Swedish and American merchants. The offices and warehouses of these European companies were known as 'factories' or 'hongs.' Trade was tightly restricted by the Chinese authorities: they conferred inferior social status upon the tradesmen, limiting their presence to the factory area outside the Old and New Cities, and making the success of their business dependent upon adherence to Chinese customs and culture. Nevertheless, trade with China was integral to the economies of European companies, and the nations they represented.

This view is the work of an unknown Chinese artist, and gives an apparently faithful rendering of Canton in the third quarter of the eighteenth century. In it the walls of the Old City are clearly visible, with the New City below, as are the Western factories lining the Pearl River. The Dutch and English flags are joined by those of the Danish and Swedish factories, with the white flag of France between them. The numerous Chinese junks in the river are owned by rivals to the European trade, but the absence of European vessels is misleading. These were not permitted into Canton harbour, having instead to moor seven miles away in Whampoa. The peaceful nature of the scene belies the rivalries and differences that existed between the Europeans themselves, and between Westerners and the Chinese authorities.

Painting in distemper, the artist has adopted Western artistic conventions to make the work more familiar to a European client. Looking closely at the city walls, and the veranda protruding from the English factory, for example, it is clear that the artist was imitating perspective, not drawing according to its geometrical principles. Yet despite such attempts to conform to Western approaches, traditional Chinese artistic traits shine through in the painting of trees and in the minute, faithful attention to detail. These, of course, were the features most attractive to those viewing the map in a European parlour.

Exactly how the view arrived in the geographical collection of George III is unclear. It was common for gifts to be exchanged between the companies and the Chinese authorities at the start of every trading season, and it is not inconceivable that the view might have been given as a gift to an East India Company official, and then passed from him to the king. It is equally possible that it may have been commissioned especially for George, who had a thirst for maps and views of parts of the world; or it could simply have been purchased by a merchant in one of the many shops which lined the bustling streets in between the factories on Canton's harbour, eventually to be bought as part of a lot at auction in London by one of the king's agents.

Conner (2009).

THE SCHOOLROOM

The schoolroom is probably the most familiar, and generally the least magnificent, setting for large wall maps. Whether in the lecture hall of a medieval university or on the walls of a classroom in a twentieth-century comprehensive, the maps have had to be clear, with strong but simple lines and blocks of colour, so that they could be seen and understood from a distance. Generally the amount of text has been restricted, since little of it could have been read from a distance and the teacher was expected to give verbal explanations of the most important features. Almost invariably the purpose of the maps was to reinforce accepted orthodoxies, whether religious – as in the case of medieval world maps or later maps showing the Holy Land – national or racial.

All too often, school maps have been of appalling quality in terms of appearance and paper, and outdated in intellectual content. In part, this was because official views of the world tend (with certain outstanding exceptions) to be to be legalistic and conservative. For over twenty years after 1945, for example, because the German Federal Republic had not signed a peace treaty with Poland recognizing the line of the rivers Oder and Neisse as forming the German-Polish border, schoolchildren in West German schools had maps showing the pre-1945 borders of Germany extending deep into modern Poland, though with the qualification that the areas were under Polish administration. The conservatism could also be due to the virtual absence of any pressure for innovation. As a result old map images retained their validity, regardless of improvements in surveying techniques and even, sometimes, changes in political realities.

This conservatism was a boon for unscrupulous, or hard-pressed, map publishers. Between about 1700 and 1820 several British and French publishers cynically saw school maps as an easy way of making money from worn-out copperplates with obsolete, and frequently incorrect, information. Sometimes the first and last appearances of such maps were separated by nearly 200 years, as was the case with a highly decorative map of the Holy Land. Originally published by John Speed in the early seventeenth century and republished by Robert Green in 1682, it was still being printed in the nineteenth century, together with a florid dedication to Charles II, even though the copperplates had had to be re-engraved following the destruction of the originals during the Great Fire of London in 1666.[114] In this case the

conservatism might have been justified by reference to the map's historical and Biblical character. In other cases the blemishes were concealed beneath titles stressing the modernity of the map and by side panels containing imaginative but badly engraved images of (for instance) earlier rulers and/or important historical events. In 1796 the London publishers Haine & Son sought to disguise the age of their image of England and Wales, dating back to 1677, by lining its edges with eight illustrations of stirring events from British history. Six of them, in line with contemporary taste for British antiquity, which also found expression in 'Strawberry Hill' gothick architecture, are episodes from Anglo-Saxon and medieval history and fable. The one more recent episode, showing Elizabeth I addressing her troops in Tilbury at the time of the Armada is misdated to 1587.[115] It is an all too typical display of the limitations of most school maps.

Those who could afford it, however, were able to purchase large, elegantly produced teaching maps of considerable sophistication and intellectual distinction – and intended for study at close quarters by no more than a handful of students and their tutor. The Dutch were pioneers of this sort of map, publishing large numbers of attractive wall maps of the world and its parts with decoration by, or based on the designs of, professional artists such as Rembrandt's teacher Pieter Lastmann or master engravers like Romeyne de Hooghe. The maps were sometimes copied from the work originally produced by surveyors and mapmakers of other countries which did not have the infrastructure to print, publish and market multi-sheet maps (pp.62–3 and 65–7).

The Dutch achievement was exceeded by the wall maps produced in eighteenth-century France. The French *Académie des Sciences* collaborated with commercial map publishers to ensure that the findings of the latest expeditions led by government-supported French geographers, explorers, scientists and mathematicians, who then led the world in such matters, were conveyed to the public in the most elegant way possible. The most fashionable artists and engravers were commissioned to ensure that these scientific maps were as visually pleasing as possible. The published maps were instructive and ornamental, were sold throughout Europe and – equally important – were admired as outstanding products of French civilization (p.154).[116]

Teacher with Students, *c*.1325. British Library, Royal Manuscript 17e 3, folio 36

Jan Miense Molenaer, *The Schoolroom*, 1634. Staatliche Kunstsammlungen, Kassel

In the nineteenth century most West European countries were able to produce handsome and instructive wall maps in large quantities. The fashion for thematic mapping added new and usually colourful variety. William Smith's geological map of Great Britain of 1815 has recently been acclaimed by Simon Winchester as one of the greatest and most significant maps ever to have been produced.[117] Von Czoernig's large ethnic map of the Habsburg Empire, dating from 1855, is even more colourful and attempts as honestly as possible to illustrate and explain the immensely complicated make-up of that supranational state.[118] By the close of the nineteenth century, chromolithography was being used to create encyclopaedic maps of the Russian Empire. Visual delights attempted to capture every aspect of the empire including its wild life, industry, the history and culture of its peoples – all were presented in a manner designed to bolster support for the tsar and traditional Russian values (pp.156–7).

From the 1820s a new, more earnest spirit – linked to the revival of geography as a subject of study in its own right – became noticeable in the quality of the maps published specifically for the normal run of schools. In Britain the rival Anglican Society for the Propagation of Christian Knowledge and the secular Society for the Diffusion of Useful Knowledge competed with each other to produce up-to-date, accurate and reasonably priced wall maps for schools. It was a pattern repeated throughout Western Europe, and in the colonies of European powers throughout the world. These maps could open the eyes of poorer pupils to the world. In the words of just one of them, Henry Batchelor, a coachman's son attending a north London 'academy' in the early 1830s, 'There were large maps in the school for the upper classes. I got some boys to explain them to me. The knowledge I gained of the map of the world and of maps of the various countries I have always regarded as the one great acquisition which I made at that school.'[119]

By the end of the nineteenth century, however, school maps and atlases were being increasingly subverted into tools of overt political propaganda, a situation that endured for most of the twentieth century. Even the most culturally and politically disturbing of maps, for instance those produced in Hitler's Germany, could be works of art, with the elegance of lettering being particularly distinguished (pp.158–9).

Computers, Geographical Information Systems and the coming of Google Earth in the twenty-first century have probably signalled the end of the wall map in schools, but they have certainly not marked the end of the magnificent artistic map. A map-like image cannot be understood without clear lettering and sensitive colouring, even on a computer screen, and they continue to act as the sugar that pupils need to help the medicine of useful information to go down.

114   The title of one of the British Library examples of the map sums up the story: *[The Land of Ca]naan [described, with] the City [of] Jerusalem. Published by Rob. Greene. This map was made and finished by Mr John More ... And after his death was published at the only charge of ... Mr John Speed. And whereas ye Plates were destroyed in the dreadfull fire 1666: At ye desire of severall learned persons, the whole was revised & corrected by a carefull hand. Reprinted at ye proper cost of Rob: Greene ... In this edition there are above 400 Errors purged out and corrected. And the workmanship of it is very much improved. [This Map was Ingraved by Francis Lamb, Ano. 1682.] To His ... Majesty Charles II ... Thi[s M]ap is humbly Dedicated and Presented by your Majest[y's] true and faithfull subject Robert Green* (London, Carrington Bowles, c.1780), British Library Maps STP.
115   British Library Maps STE.
116   Joseph W. Konvitz, *Cartography in France 1660–1848: science, engineering and statecraft* (Chicago, University of Chicago Press, 1987); Mary S. Pedley, *The commerce of cartography: making and marketing maps in eighteenth-century France and England* (Chicago, University of Chicago Press, 2005); Monique Pelletier, *Cartographie de la France et du monde de la Renaissance au Siècle des lumières* (Paris, BNF, 2001).
117   Simon Winchester. *The map that changed the world: the tale of William Smith and the birth of a science* (London, Viking, 2001).
118   Johannes Dörflinger, 'Sprachen- und Völkerkarten des mitteleuropäischen Raumes von 18. bis in die zweite Hälfte des 19. Jahrhunderts' in 4. *Kartographiehistorisches Colloquium*, Karlsruhe, 1988; *17–19.März 1988: Vorträge und Berichte*, Wolfgang Scharfe, Heinz Musall & Joachim Neumann (eds) (Berlin, Dietrich Reimer, 1990), pp.183–95; *Austria Picta*, pp.130, 149, 157, 159, 160, 163, 169, 174, 358, 370; Peter Barber, 'Putting Peoples on Paper', *The Map Book* (London, Weidenfeld & Nicolson, 2005), pp.272–3.
119   *Gin and Hell-fire: Henry Batchelor's memoirs of a working class childhood in Crouch End 1823–1837*, Peter Barber (ed.) (London, Hornsey Historical Society, 2004), p.55.

A government village school, Musgang, Punjab, 1870,
British Library 1000/46 (4703)

A class in the Parsi Virbaiji school, Karachi, 1873,
British Library 1000/46 (4655)

Teaching piety and patriotism

This map may be the earliest surviving school map. Too big to fit into a book, it never seems to have had an explanatory text – unlike the smaller world maps which resemble it in form and are to be found as integral parts of several manuscript copies of the *Polychronicon*, by the fourteenth-century Chester writer Ranulf Higden. Judging from the great number of surviving manuscript and printed copies and references to it in contemporary literature, we know that this work must have been very popular throughout Europe. The map's size, clear, dark lettering and extensive blank spaces suggest instead that it was intended to be seen from a distance, and to be explained verbally by a teacher. The extensive annotations, consisting of place names over England, with a few added on the European mainland, were inserted about twenty-five years later. The map's excellent state of preservation is probably due to the re-use of the reverse side of the vellum after 1447 as part of a family tree of the Boteler family.

The context and content of the Evesham map offer further support for the hypothesis that it was meant for purposes of instruction. It is almost identical in general form – particularly in the shapes of the Mediterranean and the Red Sea, the relative sizes of the continents, and the form of what, here, are the British Isles but on its prototype is a fantasy island in the sea encircling the Earth – to one of the earliest detailed world maps, probably created in southern France in the late eighth century (Bib. Apost. Vaticana MS Vat. Lat. 6018, fols. 63v–64). This conservatism may be ascribed to the fact that the Evesham map, far from having been made for a specific purpose, was a standard map, intended to accompany sermons and lectures and to illustrate conventional learning. Such maps must once have been relatively common. The Evesham map's resemblance to the later maps illustrating the *Polychronicon* may reflect Higden's choice of an off-the-shelf world map to illustrate the geographical section of his text.

The Evesham map reinforces the accepted medieval European world view. Adam and Eve are shown in the Garden of Eden at the top (east). As the eye travels westwards down the map it notices the Tower of Babel, the passage of the Israelites across a (very red) Red Sea, and a multi-towered Jerusalem. The names of rivers and of the Roman provinces would have enabled a skilled teacher to remind his pupils of the biblical and mythical events associated with these places.

As one continues into Europe, so time advances towards the present. We pass the ancient Greek holy shrine of Mount Olympus and then, marked with a medium-sized tower, Rome. We cross into northern Europe by the St Gotthard Pass, which had become viable only a century earlier, and reach a tower representing St Denis, the burial place of the kings of France. Its large size reflects its importance for the kings of England, who claimed to be the legitimate kings of France. The insignificant tower to its left is Paris, but Calais, which Edward III had conquered in 1348, is the large tower below. The other towers represent England's important trading partners, Bruges and Cologne.

Most striking of all, however, is an outsized England across the Channel, stretching from Scandinavia to the Mediterranean; it is separated from a small Scotland by water and from Wales by a sea-like River Severn. Beyond lies Ireland.

We are in the Europe of Henry V, and the Evesham map must have helped to instil patriotism and piety into young Englishmen more than a century before Shakespeare.

Barber (1995), pp.13– 33; College of Arms (1936), p.59, no.68

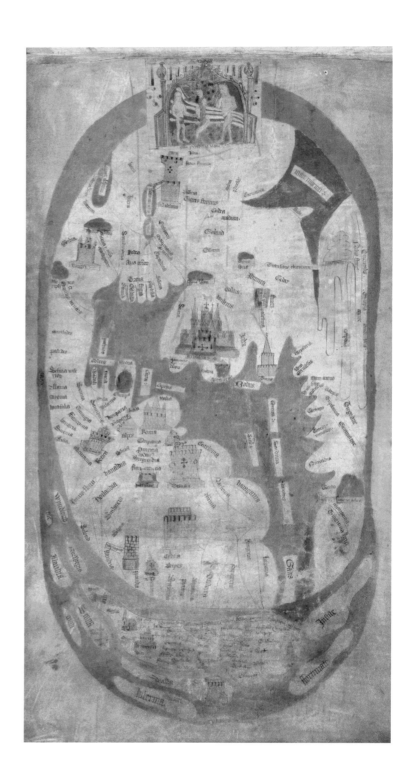

The Evesham world map, Evesham, *c*.1390–1415
Manuscript on vellum, 94 x 46 cm on sheet measuring 99 x 55 cm
College of Arms, London: Muniment Room 18/19

The Victorians in colour

This example of George Frederick Cruchley's school map of Europe, 'compiled for the use of colleges and schools', provides an alternative impression of the Victorian classroom to that conveyed through contemporary photographs. Such images often show seated rows of glum children in equally gloomy surroundings, but brightly coloured wall decorations such as maps were the norm rather than the exception, and were tailored specifically for their intended young audience. The considerable size of Crutchley's map would have enabled it to be read clearly from some distance away, and the bold dark print and legible lettering, while fulfilling the same function, would also have suited a less literate audience. So much ink has been used, in fact, that the lettering is raised up above the paper.

An increase in the number of maps made specifically for schools in Victorian Britain is attributable to two main factors. Firstly, the increasing number of schools, founded from the 1830s onwards by charitable organizations with the assistance of government money, also meant that not only were more children able to receive an education, but that there was a more lucrative market for school maps. Secondly, cheaper methods of printing, such as lithography, enabled larger quantities of maps to be produced at a smaller cost, while the advent of colour printing led to a far more streamlined production process. Cruchley's map, however, would have been something of a deluxe model. It is an engraving printed on nine sheets which have been joined together, and the colouring has been applied entirely by hand. Although certain English publishers are recorded as utilizing the artistic skills of children in workhouse schools for the laborious (though at least initially quite enjoyable) process of colouring maps, the careful application of paint in Cruchley's map suggests that he was not one of them.

The content of the map embraces numerous topics, making it an important visual aid for lessons other than simply geography. Such an approach is consistent with an increasing tendency during the nineteenth century for maps to display a wide range of social, botanical and climatic information. Yet, out of its school setting, the different types of information appear unrelated to each other. For example, the isothermal lines providing the northern and southern limits for the growing of certain species of plants ('Northern limit of the Oak' and 'limit of the region of Palms') sit awkwardly alongside the locations and dates of key European land and sea battles. The population of minor islands are provided on the map, while the populations of the major nations are given in a separate table. Unsurprisingly, England and Wales is at the top of this table, despite not being the most populous of the major nations. Equally, the single colour used for the whole of the British Isles projects a particular national bias, which doubtless would have had its equivalents in schoolrooms of other nations across Europe.

George Frederick Cruchley, *Cruchley's enlarged Map of Europe. Compiled for the use of Colleges and Schools, showing the principal physical features, etc*, London, 1851

Copperplate engraving on nine sheets, 171 x 208 cm

British Library Maps S.T.E.

A princely education

In 1741 George II's younger, and favourite, son William Augustus, Duke of Cumberland, reached the age of 20. It was time for him to be introduced to the complexities of the balance of power within Germany or, to use its legal name, the Holy Roman Empire of the German People. This was where the Hanoverian lands of his father were situated, and where William Augustus was likely to be active in the military career for which he was destined.

It was a particularly sensitive time. Following the death without male heirs in October 1740 of Emperor Charles VI, portrayed at the bottom-left of the map, the new king of Prussia, Frederick II, had invaded Silesia, unleashing a European war. A new emperor had yet to be elected – the Elector of Bavaria, who enjoyed French support, was only to be elected in February 1742 – and meanwhile the war spread.

Not for William Augustus the standard school wall map. Only a specially adapted map containing up-to-date specialist information would do. The man chosen for the task of instructing William was Augustus Schutz, a cultivated gentleman descended from a family of Hanoverian courtiers who had been living in England for several decades. As master of the robes and keeper of the privy purse, Schutz enjoyed the king's special trust and favour.

Schutz chose a copy of an up-to-date map of Germany, published in 1734 by the Amsterdam publishers Covens and Mortier, and had it specially coloured to meet his requirements. As he explained, 'this map of Germany is coloured not according to the usual division … into ten Circles [the legal administrative regions] … but according to the territories possessed by the several states … in order to give an idea how far some are more or less powerful. It being near impossible to express every thing by colouring only (there being hardly a sufficient variety of colours for that purpose)', he accompanied the map with an extensive explanatory text.

With only nine colours available to him, Schutz had to use the same colour on the map for two or more states, although to avoid confusion these were placed as far away as possible from one another. Beyond that he had to distinguish between states ruled by different branches of the same family (therefore taking the same colour), but which may have been enemies. Just one colour was available for the profusion of miniature states in Swabia and Franconia, but mostly their princes were virtually powerless. The same was true of the numerous autonomous bishoprics, though there were further complications in that several were ruled by the younger sons of leading Catholic dynasties or, like Osnabrück, they alternated between Protestant and Catholic bishops, and the map had to show this. However, lands that did not enjoy some degree of independent existence were excluded – even if they were owned by powerful aristocrats, such as the Liechtensteins or Esterhazys, or by foreign rulers (since in themselves, and however great their extent, they gave their possessors no constitutional voice in German or international affairs).

The Duke of Cumberland evidently appreciated the map and accompanying text. He kept them with him throughout a career marked by occasional triumph, particularly in Scotland, and by several military and diplomatic disasters in Germany.

Van Egmond (2009), p.409, no.33, state 1.

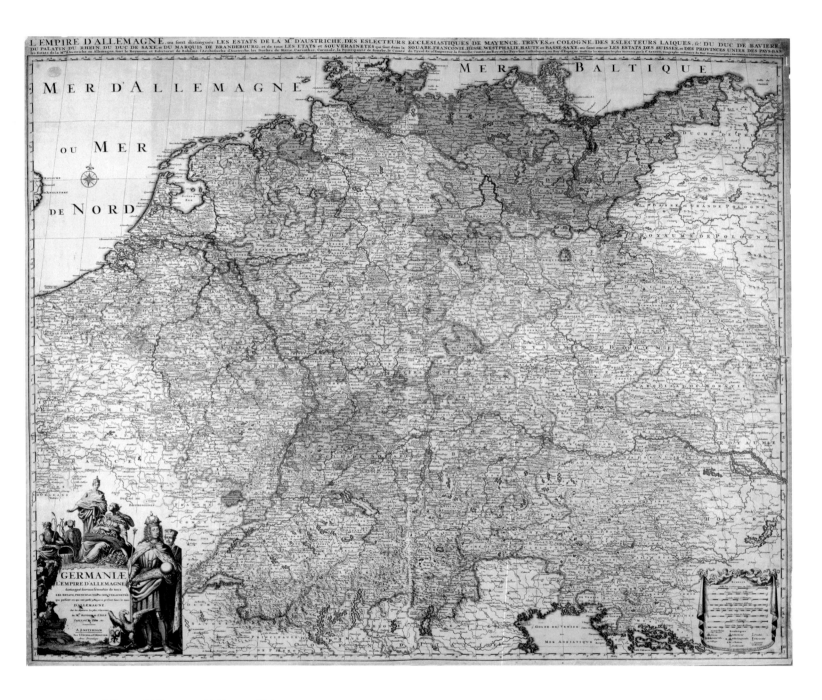

Anon., *Germaniae/L'Empire D'Allemagne* with [Augustus] Schutz,
'An Account of Germany as to the several states possessed by its different Powers
or Houses and variously & respectively Coloured in a map of Germany, which this
Accounts refers & belongs to', [1741], Amsterdam: Covens & Mortier, 1734

Copperplate engraving on four sheets and manuscript on paper,
101.5 x 119 cm (map)

British Library Maps K. Top 87.38

On its last legs

When it was first published in 1700, this four-sheet world map by Jean-Baptiste Nolin (1657–1708) was truly magnificent. Yet this example of it, printed nearly seventy years later, shows that by 1767 it had become a shadow of its former self. Over-use of a copperplate inevitably leads to wear. The once strong, dark, engraved lines and well defined and textured features of Nolin's map have become lighter, blunter and fainter after repeated printings. Although old copperplates were regularly given 'facelifts' to restore some of the original vigour of the map, or to change out-of-date details and names, it was more difficult to update the style or context of a map which, over seventy years, could suffer through changes in taste. But maps such as Nolin's continued to be produced while there remained a demand for them, and one area of the market thought particularly suitable for old maps was the school.

At first thought it may seem cynical on the part of the map publishers to supply outdated products for educational purposes. However, the financial resources of a school were not as great as those of the private connoisseur for whom the map was originally made, nor was there a requirement for cutting-edge artistry and fashion. Clearly as long as a map was not geographically incorrect, it could function perfectly well in a school setting. The geography of Nolin's map was in fact completely updated in 1708, and, with minor additions, the 1708 version continued to be broadly correct up to 1767 – conveniently just prior to James Cook's discoveries in the 1770s, which changed the face of the map forever. As late as 1742 the map was still being described by Abbé Lenglet du Fresnoy in his *Catalogue des meilleurs Cartes* as 'le trésor plus accompli que nous avons eu'.

In an educational context, the profusion of illustrations surrounding the hemispheres, the work of the painter and engraver Francois-Nicolas Bocquet (d. 1716), become more than simple embellishments. The compilation of words and pictures to recount Biblical narratives, such as The Fall of Man and the Exodus, would rather have spoken clearly to a young, and possibly illiterate, audience in much the same way as the vast pictorial schemes of the religious mappae mundi would have done four centuries earlier. Like these earlier creations, much of the imagery has precedents in earlier sources: the large and shivering bearded figure representing winter has been taken from the atlas map of the Arctic produced by Willem Blaeu in 1640, while the interpretation of the World's creation in Genesis indicates more than a passing familiarity with Michelangelo's Sistine Chapel ceiling of 1510. In fact Bocquet had spent many years in Rome and also produced engravings from frescos in the Stanza di Raffaello in the Vatican of 1508–11.

Nolin's son and successor had died in 1762, and the purchaser of his stock, Jean Francois Daumont (d. *c.*1775), must have felt there was still a market for the map – even though such decoration was out of touch with the more scientifically-minded cartography promoted by the Académie des Sciences. No changes were made to the copperplate, not even the dedication to the librarian Jean Paul Bignon (1662–1743), who had by that time been dead for twenty-four years.

Shirley (2001), pp.598–601 (no.605).

Title with dedication to the deceased librarian Jean Paul Bignon

Detail showing the creation of the firmament

OPPOSITE: Vignettes of the Garden of Eden and The Fall of Man

Jean-Baptiste Nolin, *Le Globe Terrestre Represente en Deux Plan-Hemispheres…*, Paris, 1767
Copperplate engraving on four sheets, 124 x 154 cm
British Library Maps K.Top 4.31

A lesson for the teacher

The All-Russia Exhibition of Industry and Art, celebrating the feats of engineering, architecture and technology of tsarist Russia, took place in the city of Nizhny-Novgorod in 1896. The exhibition was an expression of Russian power, funded by the government, and it drew people in their thousands from various parts of the sprawling empire. One especially prominent group of attendees were schoolteachers. Six thousand of them were to visit, many with the aid of free travel.

This school map of European Russia, produced especially for the exhibition by V. Urusov with the input of teacher associations, was awarded a prize. It is an extraordinary pictorial view of Russia, containing numerous vignettes of Eastern European life, with the Northern Lights shimmering below portraits of past and present tsars. The map was accompanied by a booklet explaining every symbol, diagram and image. A map produced for children perhaps, but one which spoke primarily to the teachers who had the task of interpreting it for their young students. To understand the map, it is necessary to know something of the Russian schoolteacher's plight, and also of the motives of the Russian state.

Russia, with a largely peasant society, had benefited over the course of the later nineteenth century from improved local government aimed at reforming rural backwardness. Of key importance in this policy was the teacher, isolated before the 1890s from his or her fellow professionals and marginalized but increasingly in touch with others through teaching associations during the final decade of the century. It was essential, however, for the Ministry of Education to retain control over what was being taught in schools during a period of increasing opposition to autocratic, reactionary rule and the antagonistic foreign policy of the government.

The map is riddled with pro-tsarist propaganda that was meant to appeal directly to the emotions of the teachers, and through them to the people they taught. The supposed abundance of Russia, a nation rich in natural resources, is emphasized through the depiction of woods, fish, wheat and fruit, while the main product of each of the seventy-eight provinces is shown proudly next to its arms. The famine of only five years earlier (1891–2), in which as many as half a million people had died, is afforded no mention in this vision of Utopia. The message of Russian unity, each province linked together by a new rail network and contributing to the national well-being, is aligned with government policy. In fact the policy of 'Russification', in which Russian values and culture were forced upon occupied countries such as Finland and Poland, as well as religious and ethnic groups such as Jews, was regarded by their intelligentsia in much the same way as ethnic cleansing is viewed today. The map was published in Warsaw, capital of Catholic Poland, and its strongly Orthodox message and Cyrillic text would have caused particular resentment among the Polish teachers who used roman script.

The map also promotes the image of tsarist benevolence. Nicholas II (1868–1918) had been crowned, two years into his reign, in 1896, the same year as the Nizhny-Novgorod exhibition. The map of 1896 celebrates the event accordingly, positioning Nicholas next to Peter the Great (1672–1725) and depicting scenes from the tsar's life, including his coronation. However, by the time the map was re-published in 1903 the scenario had changed. Seventy-one teaching societies had formed between 1895 and 1903. The All-Russia teachers' movement was created in 1903. A teachers' conference in Moscow in that same year aroused an increased militancy which was later to combine with other areas of dissatisfaction against tsarist rule and explode into violence in the revolution of 1905. The vision of Utopia had failed to convince.

Seregny (1989).

M.I. Tomasik, [*Pictorial Map of European Russia*], Warsaw, 1903
Lithograph, 218 x 194 cm
British Library Maps Roll 537

НАГЛЯДНАЯ КАРТА ЕВРОПЕЙСКОЙ РОССIИ

157

Cartographic capitulation

Rudolf Koch (1874–1934) was one of twentieth-century Germany's most important typographers and calligraphers. He sought to revive and revital-ize his country's Renaissance artistic heritage, particularly associated with fine woodcut prints and superb lettering. Much of the best of this had appeared in Nuremberg, the hometown of both Koch and Albrecht Dürer. Koch was much influenced by the work of William Morris, and he established a Renaissance-style workshop at Offenbach near Frankfurt where he lived after 1906.

Koch had long been dissatisfied with the standard maps of Germany and between 1925 and 1933 he created one more in keeping with his ideas, assisted by cartographers from the Hesse Land Registry Office and artists and engravers in his workshop. The map is put together from numerous zinc plates which give the impression of woodblocks. Koch hoped that the finished map would be hung in German schools.

In an essay written in 1933, Koch lyrically described the map as 'a picture of a beautiful land, its rivers and streams, its forests and lakes, its towns with their marvellous old emblems, a picture over which the eye can walk, like a wanderer striding through God's garden with a happy and grateful heart'. His map is adorned by quotations extolling Germany by Goethe and his contemporary, the Romantic poet Friedrich Hölderlein. The overall impression is of a modernized Renaissance-style map, influenced by the style of Philip Apian's survey of Bavaria (1568) and decorated with handsomely executed (though somewhat inconsistent and incorrect) hand-painted heraldry.

At first glance the map seems unexceptionable. It title states that it depicts Germany *und angrenzende Gebiete* ('and its neighbouring districts'), and red lines show the national borders fixed in 1919. This did

Rudolf Koch, Fritz Kredel, Richard Bender and Berthold Wolpe, *Deutschland und angrenzende Gebiete*, Leipzig, Insel-Verlag, 1937

Zinc engraving on two sheets, 120 x 163 cm

Private collection

not correspond to Koch's original intention, however. The map as originally published in February 1935 was entitled 'Germany' and showed no boundary lines, giving the impression that the Reich extended over Austria and the German-speaking areas in France, Czechoslovakia and Poland. Such a vision corresponded to the century-old concept of a 'Greater Germany', or *Grossdeutschland*, nurtured by traditionally minded German nationalists. However, Koch had gone beyond this. Though he was a pious Lutheran and had done work for the Jewish community in Offenbach, the text under the title on the original proof for the map contained a swastika and proclaimed that it had been published 'in the first year of Germany's renewal'. Adolf Hitler had been appointed German chancellor in January 1933.

The map as originally published caused Nazi Germany considerable international embarrassment, the Italian leader Benito Mussolini specifically denouncing it at the conference at Stresa in April 1935 as an example of German revanchism. As a result, the German authorities banned the map, though it became a collector's item for Nazi officials. Ironically by the time that it appeared in the revised, less internationally offensive form seen here, two of the artists most closely involved with the map, Berthold Wolpe and Fritz Kredel, had been forced into exile for political and racial reasons, although they continued to be named on the map. The map itself became outdated about a year after its appearance when Hitler's Germany annexed Austria – a change that Koch would have applauded.

The association of nostalgia for the artistic past of one's country with right-wing politics was normal in central Europe during the nineteenth and twentieth centuries, but Koch had a more radical approach, tinged with elements of the socialism that characterized William Morris. It tempts one to wonder whether if he had still been alive in the 1930s Morris might not have supported Oswald Mosley and his fascist Blackshirts. At a deeper level the map offers an example of the readiness of the Nazi regime in its early years, despite Hitler's bluster, to give way in face of united and determined international opposition to its policies.

Sarkowski (1999), pp.27–34; Cinamon (2001), particularly pp.167–9.

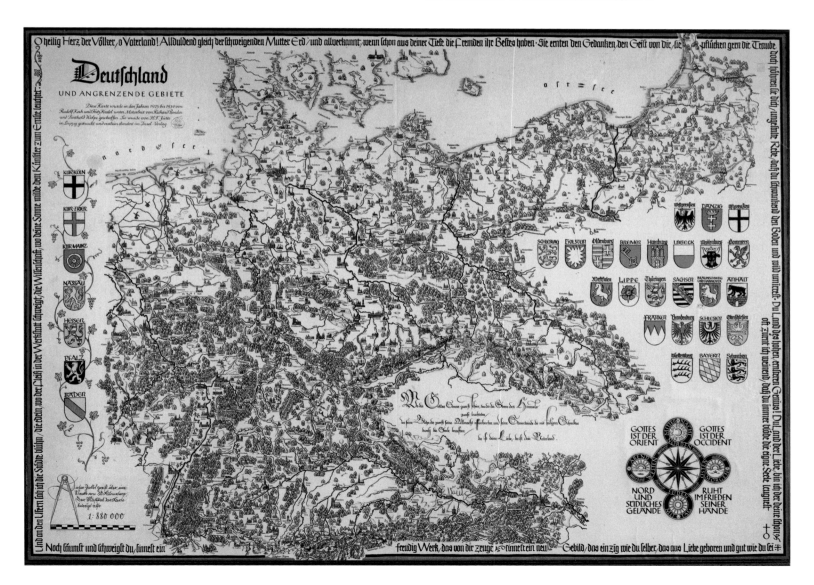

Because of the emotive connotations attaching to them and the familiarity of their outlines, map and globe images lend themselves readily for use as symbols for abstract ideas. They have been used in this way in Europe since at least the 1330s, but were to take on a new, more plebeian, life from the nineteenth century.[120]

Undoubtedly the commonest geographical form to be displayed is the globe. Before about 1800 globes were created for the ruling elites, as symbols of power and universality, but since 1850 they have been mass-produced and, while not losing their emotive power, have been utilized principally for educational purposes. Globes for display may be epitomized by the great hand-made celestial and terrestrial globes glorifying Louis XIV (measuring nearly 4 metres in diameter) that were commissioned by Cardinal d'Estrées from Vincenzo Coronelli between 1681 and 1683 for display in the palace of Versailles,[121] but there were other still bigger globes that served a more democratic clientele. In 1851, just in time to capitalize on the popularity of the Great Exhibition held in Kensington Gardens that year, an enterprising map publisher James Wyld managed to lease the run-down gardens in the middle of Leicester Square in London. He filled the space with a large building into which he squeezed a vast globe, measuring 20 metres (60 feet) in diameter. Wyld then lined the entire inner surface with a relief map of the whole world. A series of steps and platforms at various levels enabled visitors to see the world in a hitherto unimaginable level of detail.

The globe was an enormous success for the ten years that it existed. Somewhat smaller, at only 10 feet (2.030 metres) in diameter and weighing forty tons, is the Great Globe in Swanage, Dorset, made in 1887 out of fifteen segments of Portland stone. Created for a local businessman William Burt as one of a number of attractions intended to boost tourism, it still stands high on a cliff at the edge of the sea.

Globes have often been architectural features. The façades of baroque palaces were common locations, and a particularly splendid, gilded example featuring the monogram of Emperor Charles VI, is to be found over the main wing of the Imperial (now the Austrian National) Library in Vienna, built in the 1720s. Another globe, created in 1909 by René de Saint-Marceaux in the park outside the International Postal Union headquarters in Bern in Switzerland, is embraced by five Art Nouveau figures carrying letters across the world. The globe is used in a commercial context as a symbol of internationalism and international power in many large cities, from London (where it crowns the Coliseum, built as a theatre in 1902–4) to Seattle (on top of a tower block in the business quarter). A more recently created globe, appropriately rendered in black marble and tilted so that the world is viewed from the north and south poles (thus giving prominence to Africa and the Third World), forms the centrepiece of the memorial to Martin Luther King in Lake Worth in Florida.[122] Like many examples produced since the first manned space shots in the 1950s, which enabled people to view Earth from afar, the Lake Worth globe symbolizes not power or size but the fragility of the world it portrays.

Maps are also occasionally to be seen in the open air. Some serve cultural purposes, such as the large ceramic map of western Cornwall in Penzance station, which shows the major sights of Cornwall as well as the steam train that made them accessible to tourists. More often the maps are blatantly propagandistic in intent. The splendid mid-seventeenth-century Baroque façade of the church of Santa Maria del Giglio in Venice, for instance, incorporates carved plans of cities and fortresses in Italy and the Venetian Empire particularly associated with the careers and achievements of the church's patrons, the Barbaro family. In the late 1930s, aping the marble map of the Roman Empire created by Vipsanius Agrippa for the Emperor

The Great Globe at Tilly Whim, Swanage, Dorset

Plan of Candia (Iraklion) in Crete from the façade of the church of Santa Maria del Giglio in Venice, c.1680

Augustus, Benito Mussolini commissioned a series of five marble maps charting the growth of the ancient Roman Empire and culminating in a map (since removed) of what turned out to be Italy's short-lived 'Second Roman Empire' in Africa.[123] Mussolini placed the maps along the triumphal road that he rather brutally cut through the historic heart of the city, to emphasize his view that Fascist Italy was the heir of – and would surpass – the Roman Empire.

In the later nineteenth and twentieth centuries artists and publicists were not slow to realize the commercial and political potential of cartographic shapes. Posters have thus used maps to buttress a wide variety of political propaganda and commerce. Often the maps have been incorporated as an emblematic element in a larger design. Others have provided the visual framework on which a scenario is played out. It is in this context that many cartographic posters share similarities with the language of political cartoons. Crucially, however, the requirement for posters to communicate quickly with a large audience necessarily involves increased dimensions and stylistic boldness, as well as a more singular, direct message. In this way posters have often had more of an affiliation with graphic design and public art than political illustration, in accordance with the urgency of their messages. Like all display items of an ephemeral nature, they were discarded once they had served their purpose.

The lithographic printing process developed during the nineteenth century not only allowed for bold, colourful images to be produced in large numbers, but was also a relatively cheap method of communicating with a mass audience. There seems to have been a whole school of designers who from 1877 utilized the idea of an octopus spreading across the map as a means of whipping up hostility to whoever happened to be the enemy (p.164). During the early 1920s some leading Soviet graphic designers incorporated cartographic elements, into their posters to glorify the achievements of the communist worker and to vilify the bolsheviks' opponents (p.167).

The same techniques could, by contrast, be used to associate big business with patriotism, as in the case of Imperial and International Communications Ltd (p.167) – or to persuade consumers to take cruises, travel by train, fly by certain airlines, drink tea (p.166) or purchase fizzy drinks – often adding a touch of humour in the process.[124] The effectiveness of all such posters depends on the skill with which the map is integrated within the overall design, the interplay between the figurative and cartographic elements, the clarity of the lettering and the balance of colours needed to give a striking overall impression.

120 Gillian Hill, *Cartographical Curiosities* (London, British Library, 1978), pp.39–49; Guy Roux and Muriel Lahaine, *Art et folie au Moyen Age: aventures et énigmes d'Opicius de Canistris (1296–vers 1351)*, (Paris, Léopard d'Or, 1997).
121 Monique Pelletier, '*Des Globes pour le Roi-Soleil. Les origins des "globes de marly"*; and '*Le Cardinal, Le Moine, Le Roi et les autres. Les acteurs des globes du Roi-Soleil*' in *Tours et Contours de la Terre. Itinéraires d'une femme au coeur de la cartographie* (Paris, Presses de l'école nationale des Ponts et Chaussées, 1999), pp.22–23, 33–45. Because of their size, the display of the globes has proved to be a headache. They were never displayed at Versailles and are now accommodated in the modern Bibiothèque nationale de France building at Tolbiac in Paris.
122 See also Jan Mokre, '*Globen unter freiem Himmel: Beispiele aus Wien*', *Der Globusfreund* 47/48 (1999/2000), pp.125–41.
123 Heather Hyde Minor, 'Mapping Mussolini: Ritual and Cartography in Public Art during the Second Roman Empire', *Imago Mundi* 51 (1999), pp.147–62.
124 Heather Child, *Decorative Maps* (London, Studio Productions, 1956), pp.24–5 (which reproduces a map of 'Schwepperhire').

An Art Nouveau globe at the International Postal Union, Berne, Switzerland

Marble maps of the Roman Empire, via dei Fori Imperiali, Rome

A British general election

On 8 March 1880 the British prime minister, Benjamin Disraeli, Earl of Beaconsfield, called a general election. These two posters, commissioned by a leading map publishing firm G.W. Bacon, present alternative views of his situation and achievements. Both are probably by Fred W. Rose, whose 'octopus' maps had caused a sensation three years earlier.

In the first poster Disraeli appears as a patriotic hero, dealing a death blow to the Liberal leader Lord Hartington. Disraeli's principal opponent William Gladstone, in contrast, is shown as a Scottish windbag, opportunistically profiting from current discontents both at home and abroad. The cannon refers to Gladstone's criticism of the unpopular war in Afghanistan which had broken out in 1878, while his policy of granting home rule to Ireland is portrayed as a devilish plot to break up the United Kingdom by giving Ireland the means to sever bonds to England. The 'post cards' are probably a reference to Gladstone's mobilization of public opinion through the use of what today would be called mailshots, as well as through the power of the speeches that he gave during the Midlothian campaign the previous November. Ireland, suffering from distress and famine, is being led astray by an evil adviser representing the Irish nationalists, though the union with England actually 'aids her … materially to support her heavy burden'.

In the second cartoon there is much play on Disraeli's vanity. 'His Imperial Majesty King Jingo I' (a play on the words of a bellicose popular song) is shown being worshipped by his subservient cabinet, but about to be toppled by 'public opinion' operating though the ballot box. The barrels with 'X' scratched on them suggest that voting papers would have the same dramatic effect as the gunpowder that Guy Fawkes had hoped to use in 1605. Disraeli's campaign banner, already broken, proclaims 'Peace! With Honours'. This is an ironic play on his words of two years earlier. After having apparently saved his country from involvement in a European war, Disraeli had proclaimed that he had secured 'peace with honour' – in gratitude for which Queen Victoria had made Disraeli an earl. In 1876 Disraeli had created the Queen 'Empress of India' by way of the Royal Titles Act which is also referred to on the poster. The political situation this poster portrays could hardly be grimmer. Economic depression, immense public debt, enormous deficits and rising bills (perhaps a reference to the cost of constructing public sewers) is compounded by resentful tenant farmers, angry at broken promises, and Irish indignation fuelled by famine and evictions. The only solution is William Gladstone, who delivers the fatal wound through an oversized pen used to champion 'Truth and Right'. Holding a banner summarizing his programme, against a background of 'enlightened public opinion', Gladstone treads underfoot the worst manifestations of Disraeli's policies, which he implies were entirely self-interested.

The electorate were to take Gladstone's side, and a Liberal government was returned on 28 April.

Fred W. Rose, *Comic Map of the British Isles indicating the Political Situation in 1880*, London, G.W. Bacon, 1880

Lithograph, 64 x 51 cm

British Library Maps 1078. (45.)

'Nemesis' [Fred W. Rose?], *The overthrow of His Imperial Majesty King Jingo I: A Map of the Political Situation in 1880 by Nemesis*, London, G.W. Bacon, 1880

Lithograph, 69.5 x 52.5 cm

British Library Maps 1078. (44.)

European peace – on France's terms

This poster, probably dating from 1869, offers a simple way of transforming international enmities, disquiet, mistrust, deception, ruin and war into humankind's dream of peace, reconciliation, confidence, prosperity, plenty, glory, science and wealth. The big question mark surrounding Europe becomes an air balloon of hope. All that is needed is to remove the trouble spots identified in black in the first map. These were the lands around Rome that the papacy continued to rule after the unification of the rest of Italy in 1859–60, the Rhineland, the province of Schleswig and the province of Prussia. The first would enter into some sort of loose federation with the rest of Italy while maintaining its essential independence, the Rhineland would be annexed to France, Schleswig would again become part of Denmark and Prussia would be ceded to Russia. What simple sacrifices in return for the blessings of peace!

The acute observer might observe that France itself would not be obliged to make any concessions to achieve this Utopia. Far from it – the changes would have resolved most of the nightmares of the increasingly embattled French government under Napoleon III. His continuing promise to maintain the pope as a territorial ruler infuriated the new rulers of Italy, destroying any credit that would otherwise have been his due through France's vital diplomatic and military support for Italian unification in 1859–60. The cessions of German lands would fatally weaken France's arch-enemy, the kingdom of Prussia, and serve to transfer Germany's industrial heartland to France. Prussia had become the dominant power in Germany in 1866 following its crushing defeat of Austria, which had then been excluded from the new German Confederation. Napoleon III feared a Prussian invasion of France with good reason.

Within a few years peace had indeed been achieved – but France had paid the price. After Napoleon allowed himself to be provoked into declaring war on Prussia in 1870, German forces captured Paris, a new German empire was proclaimed in Versailles and German-speaking Alsace was ceded to Germany. Meanwhile the Italians took advantage of French weakness to conquer Rome and, in the wake of invasion and defeat, France was plagued by civil war. In one respect the poster was truthful: the French vision for the future, like the balloon depicted on it, proved to be all hot air.

P.M. Barber, 'A European Union' (2005), pp.278–9.

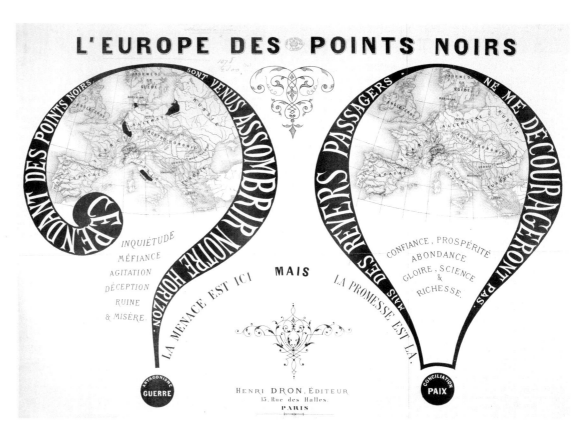

Henri Dron, *L'Europe des Points Noirs*, Paris, Henri Dron, *c.*1869
Lithograph, 53.5 x 71.5 cm
British Library Maps 1078. (26.)

Russian, Prussian and Churchillian octopuses

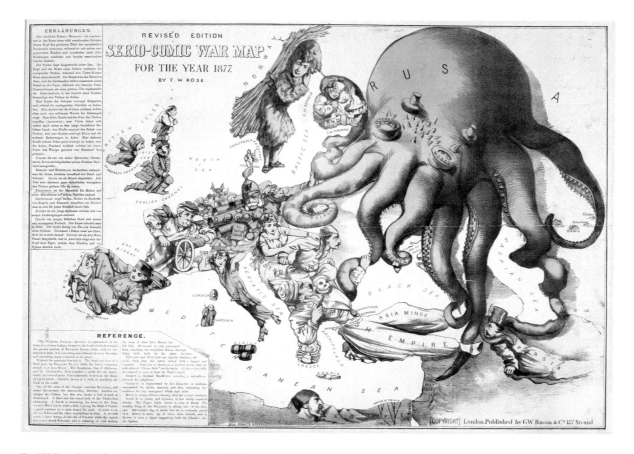

Fred W. Rose, *Serio-Comic War Map for the year 1877.*
Revised edition, London, G.W. Bacon, 1877

Lithograph, 55.5 x 71 cm

British Library Maps *1078. (45.)

Once Fred W. Rose had created the 'Octopus' map of Europe, it proved difficult to rid propaganda maps of them. They first appeared in the autumn of 1877, when Europe was poised on the brink of war. This map, a revised version of Rose's original, peddles a message of distrust of Russia that continues to offend many Russians to this day.

Russia threatened to invade the enfeebled Ottoman Empire in support of its fellow Christian Bulgarians who had been the victims of a Turkish massacre (indicated by a skull). Britain and Germany were determined that Russia should not conquer Constantinople and thereby gain direct access for its fleet to the Mediterranean and the Middle East.

Rose shows Russia as an octopus with two eyes (representing St Petersburg and Moscow). It was throttling Poland and nearly strangling Finland (both of which were then part of the Russian Empire (p.157), while its tentacles threaten the shah of Persia, central Asia, Armenia, the Holy

Land and Constantinople, shown as the Sultan's gold watch. Greece, portrayed as a crab, is ready to join the Russian attack on Turkey. The old German Emperor, Wilhelm I, tries to push back the octopus, and Hungary wants to intervene but is restrained by Austria, while England and Scotland look on anxiously. The other European countries, however, are primarily absorbed by their own affairs, raising the prospect that Russia might be able to get away with its aggression. In the event Germany and the United Kingdom made common cause and, following a conference in Berlin, prime minister Disraeli was to bring home that 'Peace with Honour' subsequently satirized by Rose.

The second map, published in Paris during the First World War, depicts Prussian-dominated imperial Germany as an octopus, spreading its tentacles throughout the continent. The map of Germany illustrates Prussia's expansion since 1871, with special emphasis on its acquisition of

164

Maurice Neumont, *La Guerre est l'Industrie Nationale
de la Prusses* ['War is the national industry of Prussia'], Paris,
*Conférence au Village contre la propagande ennemie
en France*, December 1917

Lithograph, 60 x 80 cm

British Library Maps CC.5.a.547

S.P.K., *Confiance – ses amputations se poursuivent
méthodiquement* ['Be reassured – the amputations (of his
tentacles) are proceeding methodically'], Paris, 1942(?)

Lithograph, 119 x 80.3 cm

British Library Maps CC.5.a.546

Alsace from France in 1871 and the parts of northern France conquered by
Germany in 1914. The growth of Prussian militarism is represented in the
differing sizes of the Prussian soldiers in the upper-right corner. They range
from the minute Prussian infantryman of 1715 to the massive German
infantryman in full battle-dress uniform dated 1914.

Quotations drive the point home. The title, a quotation from the
French revolutionary Mirabeau in 1788, defiantly states that warfare is
Prussia's national industry, while at the bottom a quotation from Marshal
Pétain no less defiantly declares that 'attacked, we can do nothing but
defend ourselves in the name of Liberty and ensure our own survival'
(June 1917). Parallels are drawn between Alsace–Lorraine in 1870–1 and
the invaded French *departements* captured during the German offensive
of 1914. On a more sinister note the poster quotes the *Pandeutscher
Verband* (All-German League) as threatening shortly before the outbreak

of war that the German people should arise and become masters of the
inferior peoples of Europe.

In the third poster, probably published in Vichy France and
intended to sustain its citizens' morale, a demonic green-faced, red-lipped,
cigar-smoking Winston Churchill has become the octopus. But his attempts
to seize Africa and the Middle East are being thwarted by the Axis forces,
who cut his tentacles so that they bleed profusely. The names on the map
adjacent to the amputated tentacles include Norway, Germany, Syria,
Somalia and Libya, all referring to British political and military set-backs in
the period 1940–1. Mers el Kebir in North Africa and Dakar in West Africa
receive special mention because British intervention there had left many
French dead. As the title of the map explains, the citizens of Vichy France
should have confidence in the power of the Axis which is methodically
amputating the tentacles of Churchill and the Allies.

Bolsheviks and big business

During the early years of Communist Russia, lithography was utilized to great effect by progressive artists working to support and defend their new nation and ideals. In *The Don Basin: Heart of Russia* of 1922, the Eastern Ukraine coal mining area is likened to a large red heart with arteries fuelling the many factories, emphasizing the importance of the region to Russian industry, with two muscular miners standing on either side of the title. In *The Hunger Spider* of 1922 by Dmitri Moor (1883–1946), it is suggested that the dishonestly accrued wealth of churches, mosques and synagogues be returned to the Ukraine, in order to save those starving during the famine of 1922–3. The complicity of the religious denominations and the harshness of the situation are strongly emphasized in the stark, two-colour printing and jagged design. The sinister use of the spider recalls earlier uses of creatures such as the octopus in political cartographic cartoons to highlight the evil qualities of the aggressor.

Moor's poster, produced in 1921 and entitled *Be on Your Guard*, is more satirical in comparison. Here, Eastern Europe is set as the battle-ground with the monumental red soldier urged to exercise vigilance against encroachments into Communist Russia's borders. Personifications of bourgeois capitalism are shown waiting in the Baltic States and Poland, the latter of which was being surreptitiously backed by France, while the text quotes from the Bolshevik commissar for war, Leon Trotsky. The size of the figure of the Red Guard was designed to be as much an embodiment of the Soviet state and its soldiery as a deterrent to neighbouring nations; it is the sort of visual device that has been used on maps of theatres of war since at least the sixteenth century. With such overt messages, the huge discrepancy of scale becomes part of the map's function, not an aspect of its shortcomings as a cartographic object.

The motivation for Russian artists such as Moor was implicitly the security, well-being and unity of the Soviet state, yet the language they employed could be just as effective for the expression of opposing ideologies. British advertising posters produced from the 1930s by A.E. Halliwell show business interests intertwined with imperialism and empire. In the face of competition from overseas radio communications companies, a number of British telegraph companies had joined to form Imperial and International Communications Ltd in 1928 (the company became Cable & Wireless Ltd in 1934). Clearly the safeguarding of the British Empire depended on Britain retaining control of the means of contact with its colonies across the globe, and the poster *All British Routes* of 1933 combines the world map with a lion and Union Jack in a way that not only emphasizes the dominant place of Britain, but does so in an incisive – and modern – visual manner. It is ironic that such a futuristic poster style should be employed for a form of communication which was fast becoming obsolete.

The global image of Britain is evoked once more in MacDonald Gill's poster *Tea Revives the World* (see overleaf), issued by the International Tea Market Expansion Board in 1940. A pictorial world map is covered with images and quotations detailing the importance of tea in history and culture. The message is a celebration of Allied economical superiority, a rallying cry in time of war, spoken through a history Britain's adopted national beverage. It also underlines the mutual compatability of business and national interests, for the foundation of much of Britain's wealth was the East India Company trade in tea with Ceylon and China. MacDonald Gill, brother of the artist Eric Gill, produced a great many advertisements which made use of maps, melding a striking, colourful, pictorial style with often extensive passages of text. Boldness and information, within an instantly recognizable cartographic framework, offered real scope for the purposes of advertising and propaganda.

Dmitri Moor, *Be on Your Guard*, 1921
Lithograph, 103 x 71 cm
British Library Maps CC.5.a.545

Dmitri Moor, *The Hunger Spider*, 1922
Lithograph, 93 x 62 cm
British Library Maps CC.5.a.543

Anon, *The Don Basin: Heart of Russia*, 1922
Lithograph, 69 x 52 cm
British Library Maps CC.5.a.544

A.E. Halliwell, *All British Routes*, 1933
Lithograph, 78 x 50 cm
British Library Maps CC.5.a.548

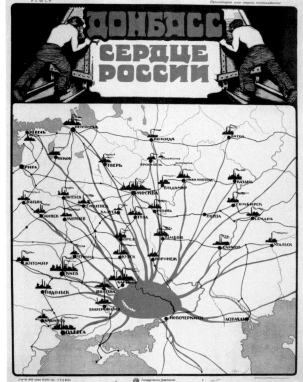

MacDonald Gill, *Tea Revives the World*, 1940
Lithograph, 75 x 153 cm
British Library Maps CC.5.a.458

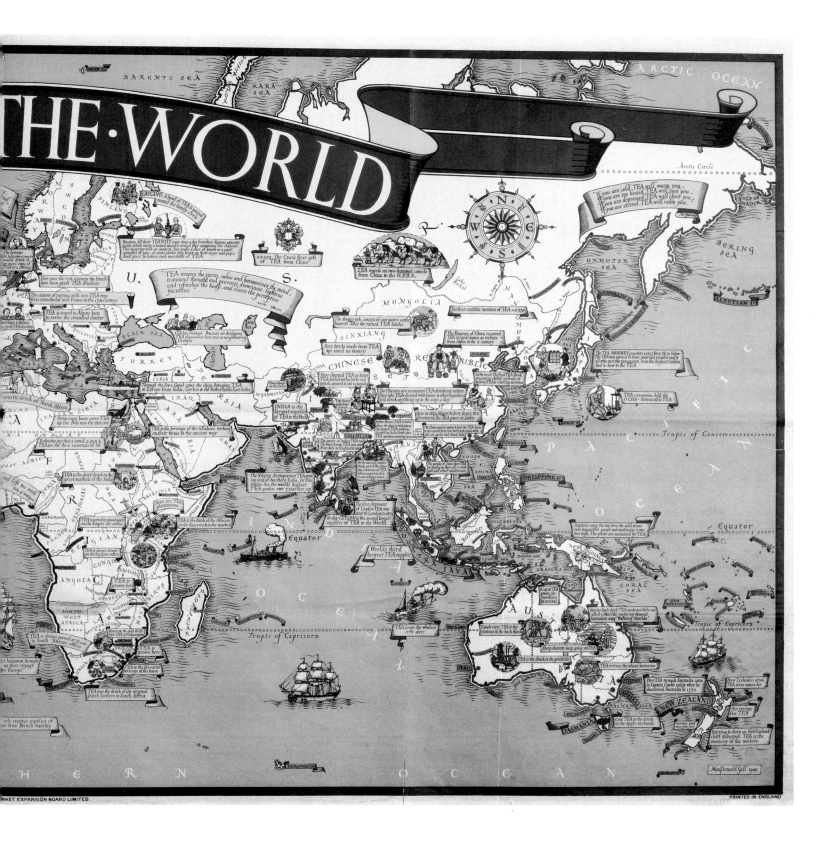

BIBLIOGRAPHY

NB The bibliography consists of every text referred to, both in the chapters and in discussion of individual maps. The titles of references in chapters are given in full footnotes on their first mention in the text. References for individual maps are have been abbreviated in the text (eg Barber 1999).

Nicholas Alfrey 'Landscape and the Ordnance survey, 1795–1820' in Nicholas Alfrey and Stephen Daniels (eds), *Mapping the Landscape. Essays on art and cartography* (Nottingham, Castle Museum, 1990), pp.23–7.

Roberto Almagià, 'On the Cartographic Work of Francesco Rosselli' in *Imago Mundi* 8 (1951), pp.27–34.

Carolyn Anderson, 'Constructing the Military Landscape: The Board of Ordnance Maps and Plans of Scotland *c.*1689–1760' (University of Edinburgh, unpublished Ph.D. thesis, 2010).

J.H. Andrews, *Shapes of Ireland: Maps and their Makers, 1564–1839* (Dublin, Geography Publications, 1997).

Jean B. Archer, 'Henry Pelham's Lost Grand Jury map of Kerry (*c.*1800): A Newly Found Derivative' in *Imago Mundi* 58:2 (2006), pp.183–97.

Germaine Aujac, 'The foundations of theoretical cartography in archaic and classical Greece' in J.B. Harley and David Woodward (eds), *The History of Cartography volume one. Cartography in Prehistoric, Ancient and Medieval Europe and the Mediterranean*, (Chicago, University of Chicago Press, 1987), pp.130–47.

Leo Bagrow, 'Old Inventories of Maps', *Imago Mundi* v, (1948), pp.18–20.

P.M. Barber, 'The Christian Knight, the Most Christian King and the rulers of darkness' in *The Map Collector* 52, (autumn 1990), pp.8–13.

P.M. Barber, 'Necessary and Ornamental: Map Use in England under the Later Stuarts 1660–1714' in *Eighteenth Century Life* 14/3 (1990), pp.1–28.

P.M. Barber and Michelle Brown, 'The Aslake Map' in *Imago Mundi* 44 (1992), pp.24–44.

P.M. Barber, 'England I: Pageantry, Defense and Government: Maps at Court to 1550' in David Buisseret (ed.), *Monarchs Ministers and Maps. The emergence of cartography as a tool of government in early modern Europe*, ed. (Chicago, University of Chicago Press, 1992), pp.26–56.

P.M. Barber, 'The Evesham World Map: a Late Medieval English View of God and the World' in *Imago Mundi* 47 (1995), pp.13–33.

P.M. Barber, 'Beyond Geography: Globes on Medals 1440–1998' in *Der Globusfreund* 47/48 (1999), pp.53–88.

P.M. Barber, 'The Copperplate Map in Context' in Ann Saunders and John Schofield (eds), *Tudor London: a map and an view*, (London, London Topographical Society, 2001), pp.16–32.

P.M. Barber (ed.), *Gin and Hell-fire: Henry Batchelor's memoirs of a working class childhood in Crouch End 1823–1837* (London, Hornsey Historical Society, 2004).

P.M. Barber (ed.), 'A European Union' in *The Map Book* (London, Weidenfeld & Nicolson, 2005), pp.278–9.

P.M. Barber, 'John Darby's Map of the Parish of Smallburgh in Norfolk, 1582' in *Imago Mundi* 57 (2005), pp.55–8.

P.M. Barber (ed.), 'Profitable and Useful, Ornamental and Allegorical', in *The Map Book* (London, Weidenfeld & Nicolson, 2005), pp.160–1.

P.M. Barber (ed.), 'Putting Peoples on Paper' in *The Map Book* (London, Weidenfeld & Nicolson, 2005), pp.272–3.

P.M. Barber, *The Queen Mary Atlas – Commentary* (London, Folio Society, 2005).

P.M. Barber, 'Mapmaking in England, *c.*1470–1650' in David Woodward (ed.), *The History of Cartography volume three. Cartography in the European Renaissance part two* (London and Chicago, University of Chicago Press, 2007), pp.1589–1669.

P.M. Barber, *King Henry's Map of the British Isles. Commentary* (London, Folio Society, 2009).

Nik Barlo, Hanae Komachi, Henning Queren, *Herren-hausen Gardens* (Rostock, Hinstorff, 2008).

Bayerische Staatsbibliothek, *Cartographia Bavariae: Bayern im Bild der Karte Bayerische Staatsbibliothek, Ausstellung, 17. Mai bis 29. Juli 1988* (Weissenhorn in Bayern, Konrad, 1988).

Geoffrey Beard (ed.), *Upholsterers and Interior Furnishing* (Newhaven and London, Yale University Press, 1997).

Guy de la Bédoyère (ed.), *The Diary of John Evelyn* (London, Headstart History, 1994).

Fathi Bejaqui, 'Iles et villes de la Méditerranée sur une mosaïque d'Ammaedara (Haïdra, Tnisie)' in *Académie des Inscriptions & Belles-Lettres. Comptes Rendus des Séances de l'Année 1997 juillet-octobre* (Paris, 1998), pp.825–860.

A. Sarah Bendall (ed.), *A Dictionary of Land Surveyors and Local Mapmakers of Great Britain and Ireland 1530–1850*, 2nd edition (London, British Library, 1997).

A. Sarah Bendall, 'Pride of Ownership' in Peter Barber and Christopher Board (eds), *Tales from the Map Room* (London, BBC, 1994), p.94.

A. Sarah Bendall, *Maps, Land and Society: A History, with a carto-bibliography, of Cambridgeshire Estate Plans c.1600–1836* (Cambridge, Cambridge University Press, 1992).

Paul Binski, *The Painted Chamber at Westminster* (London, Society of Antiquaries, 1986).

Daniel Birkholz, *The King's Two Maps. Cartography and Culture in Thirteenth-Century England* (London, Routledge, 2004).

Jeanette Black, *Commentary* on *The Blathwayt Atlas. A collection of forty-eight manuscript and printed Maps of the seventeenth century relating to the British Overseas Empire in that era, brought together about 1683 for the use of the Lords of Trade and Plantations by William Blathwayt, Secretary* (Providence, Brown University, 1975).

John Bonehill and Stephen Daniels (eds), *Paul Sandby. Picturing Britain* (London, Royal Academy of Arts, 2010).

E. van den Boogart (ed.), *Zo wijd de wereld strekt: Tentoonstelling naar aanleiding van de 300ste sterfdag van Johan Maurits van Nassau-Siegen op 20 december 1979* (exhibition catalogue: Mauritshuis, The Hague, 1979).

Molly Bourne, 'Francesco II Gonzaga and Maps as Palace Decorations in Renaissance Mantua' in *Imago Mundi* 51 (1999), pp.51–81.

Jean Boutier, *Les Plans de Paris: des origines (1493) à la fin du XVIIIe Siècle étude, carto-bibliographie et catalogue collectif* (Paris, Bibliothèque nationale de France, 2002).

William Brenchley Rye, *England as seen by Foreigners in the days of Elizabeth and James the First* (London, John Russell Smith, 1865).

Jerry Brotton, *Trading Territories. Mapping the early modern world* (London, Reaktion, 1997).

David Buisseret, *The Mapmakers' Quest: depicting new worlds in Renaissance Europe* (Oxford, Oxford University Press, 2003).

Philip D. Burden, *The Mapping of North America: a list of printed maps* (Rickmansworth, Raleigh Publications, 1996).

Philip D. Burden, *The Mapping of America II: a list of printed maps 1671–1700* (Rickmansworth, Raleigh Publications, 2007).

Tony Campbell, 'The Drapers' Company and its school of seventeenth-century chart-makers' in Helen Wallis and Sarah Tyacke (eds), *My Head is a Map: essays & memoirs in honour of R.V. Tooley* (London, Francis Edwards and Carta Press, 1973), pp.81–106.

Emanuela Casti, 'State, Cartography and Territory in Renaissance Veneto and Lombardy' in David Woodward (ed.), *The History of Cartography volume three. Cartography in the European Renaissance part one* (Chicago, University of Chicago Press, 2007), pp.874–908.

Angelo Cattaneo, *Fra Mauro's World Map and Fifteenth-century Venetian Culture* (Brepols, Turnhout, 2010).

M.C.A. de Challaye, *Notice sur la grande carte manuscrite, faite a Arques en 1550, par Pierre Desceliers, pour S.M. le Roy de France Henry II* in *Bulletin de la Société de Geographie* 4/4 (1852).

Heather Child, *Decorative Maps* (London, Studio Productions, 1956).

Jessica Christian, 'Paul Sandby and the Military Survey of Scotland' in Nicholas Alfrey and Stephen Daniels (eds), *Mapping the Landscape. Essays on art and cartography* (Nottingham, Castle Museum, 1990), pp.18–22.

Claudia Cieri Via, *'Galaria sive loggia': modelli storici e funzionali fra collezionismo e ricerca',* introduction to Italian translation of W. Prinz, *Galleria* (Modena, Panini, 1988).

Gerald Cinamon, *Rudolf Koch. Letterer, Type Designer, Teacher* (London, British Library and Old Knoll Press, 2001).

College of Arms, London, *Heralds' Commemorative Exhibition 1484–1934 Held at the College of Arms. Enlarged and Illustrated Catalogue* (London, 1936).

I. Collijn, *Magnus Gabriel de la Gardie's Samling af Äldre Stadsvyer och Historiska Planscher: 1. Kungl. Biblioteket* (Stockholm, 1915).

Howard Colvin (ed.) *et al., History of the King's Works* iii/2 (London, Stationery Office, 1982).

Howard Colvin and Susan Foister, *The Panorama of London circa 1544 by Anthonis van den Wyngaerde* (London, London Topographical Society, 1996).

Patrick Conner, *The Hongs of Canton: Western Merchants in South China 1700–1900, as Seen in Chinese Export Paintings* (London, English Art Books, 2009).

A. Cortesao and L. Texeira, *Portugaliae Monumenta Cartographica* (Lisbon, 1960).

Paolo Cortesi, *De Cardinalatu* (Castel Coresiano, 1510).

Patrick Gautier Dalché, 'The Reception of Ptolemy's *Geographia* (end of the fourteenth century to the beginning of the sixteenth century) in David Woodward (ed.), *The History of Cartography volume three. Cartography in the European Renaissance* (Chicago, University of Chicago Press, 2007), pp.285–364.

Surekha Davies, 'Representations of Amerinidians on European Maps and the Construction of Ethnographic Knowledge' (University of London, Ph.D. thesis, 2008). Summary in *Imago Mundi* 61 (2009), pp.126–7.

Elly Dekker and Peter van der Krogt, *Globes from the Western World* (London, Zwemmer, 1993).

Catherine Delano-Smith and Elizabeth Morley-Ingram, *Maps in Bibles 1500–1600: an Illustrated Catalogue* (Geneva, Librairie Droz S.A., 1991).

B. Denvir, *From the Middle Ages to the Stuarts: Art, Design and Society before 1689* (London, Longman, 1988).

Jean Dérens, 'Les Plans Généraux de Paris au xvie siècle' in Michel Le Möel (ed.), *Paris a vol d'oiseau* (Paris, Délégation à l'action artistique de la ville de Paris, 2000), pp.24–47.

O.A.W. Dilke, *Greek and Roman Maps* (London, Thames & Hudson, 1985).

Johannes Dörflinger, *'Sprachen- und Völkerkarten des mitteleuropäischen Raumes von 18. bis in die zweite Hälfte des 19. Jahrhunderts'* in Wolfgang Scharfe, Heinz Musall and Joachim Neumann (eds), *4. Kartographiehistorisches Colloquium Karlsruhe 1988 17.–19. März 1988: Vorträge und Berichte* (Berlin, Dietrich Reimer, 1990), pp.183–95.

Eamon Duffy, *Voices of Morebath. Reformation & Rebellion in an English Village* (New Haven and London, Yale University Press, 2001).

Matthew Edney, 'Bringing India to Hand: Mapping an Empire, Denying Space' in Felicity A. Nussbaum, *The Global Eighteenth Century* (Baltimore and London, The Johns Hopkins University Press, 2003), pp.65–78.

Evelyn Edson, *The World Map 1300–1492. The Persistence of Tradition and Transformation* (Baltimore, The Johns Hopkins University Press, 2007).

Marco van Egmond, *Covens and Mortier: A Map publishing house in Amsterdam 1685–1866* (G.H. Houten, The Netherlands, 2009).

Herbert Eiden and Franz Irsigler, 'Environs and Hinterland: Cologne and Nuremberg in the Later Middle Ages' in James A. Galloway (ed.), *Trade, Urban Hinterlands and Market Integration c.1300–1600* (London, Centre for Metropolitan History, Working Papers Series no. 3, 2000).

Felicia M. Else, 'Controlling the Waters of Granducal Florence: A New Look at Stefano Bonsignori's View of the City (1584)' in *Imago Mundi* 61 (2009), pp.168–85.

Brigitte Englisch, *Ordo Orbis Terrae. Die Weltsicht in den Mappae Mundi des frühen und hohen Mittelalters* (Berlin, Akademie Verlag, 2002).

R.J.W. Evans, 'The Habsburg Monarchy and Bohemia, 1526–1848' in *Austria, Hungary and the Habsburgs: Central Europe c.1683–1867* (Oxford, Oxford University Press, 2006), pp.75–98.

Ifor M. Evans and Heather Lawrence, *Christopher Saxton: Elizabethan Map-maker* (Wakefield, Wakefield Historical Publications and the Holland Press, 1979).

Francesco Fiorani, *The Marvel of Maps. Art, Cartography and Politics in Renaissance Italy* (New Haven and London, Yale University Press, 2005).

Francesca Fiorani, 'Cycles of Painted Maps in the Renaissance' in David Woodward (ed.), *The History of Cartography volume three. Cartography in the European Renaissance part one* (Chicago, University of Chicago Press, 2007), pp.804–30.

David Fletcher, *The Emergence of Estate Maps: Christchurch Oxford, 1600–1840* (Oxford, Oxford University Press, 1995).

Celina Fox, 'George III and the Royal Navy' in Jonathan Marsden (ed.), *The Wisdom of King George III*, (London, Royal Collection Publications, 2005), pp.303–7.

M. Geisberg, *The German Single-Leaf Woodcut 1500–1550* (rev. Walter L. Strauss, New York, 1974).

Alfred Haas, *Die grosse Lubinsche Karte von Pommern aus dem Jahre 1618* (Lüneburg, 1980).
P.D.A. Harvey, *The History of Topographical Maps* (London, Thames & Hudson, 1980).

P.D.A. Harvey, 'Local and Regional Cartography in medieval Europe' in *The History of Cartography volume one. Cartography in pre-historic ancient and medieval Europe and the Mediterranean* (Chicago and London, University of Chicago Press, 1987), pp.464–501.

P.D.A. Harvey, *The Hereford World Map. Medieval world maps and their context* (London, British Library, 2006).

Graham Haslam, 'The Duchy of Cornwall Map Fragment' in Monique Pelletier (ed), *Géographie du Monde au Moyen Age et à la Renaissance* (Paris, 1989), pp.33–44.

Ragnhild Hatton, *George I. Elector and King* (London, Thames & Hudson, 1978).

E. Hawkins, A. Franks and H. Grueber, *Medallic Illustrations of the History of Great Britain and Ireland to the Death of George II*, 2 vols. (London, British Museum, 1885).

Gillian Hill, *Cartographical Curiosities* (London, British Library, 1978).

D. Hodson, *County Atlases of the British Isles. Published after 1703: A Bibliography*, 3 vols. (Welwyn, Tewyn Press and London, British Library, 1984, 1989, 1997).

Hendrik J. Horn, *Jan Cornelisz Vermeyen: Painter of Charles V and his Conquest of Tunis* (Doornspijk, Davaco, 1989).

Hendrik J. Horn, *Der Kriegszug Kaiser Karls V gegen Tunis. Kartons und Tapisserien*, Wilfried Seipel (ed.), (Vienna, Kunsthistorisches Museum, 2000).

Ralph Hyde, introductory notes to the facsimile of *London &c actually surveyed* (Lympne Castle, Harry Margary in association with the Guildhall Library, 1977).

Heather Hyde Minor, 'Mapping Mussolini: Ritual and Cartography in Public Art during the Second Roman Empire' in *Imago Mundi* 51 (1999), pp.147–62.

Donald Japes, *William Payne – a Plymouth experience* (Exeter, Royal Albert Memorial Museum, 1992).

M.P. Jones, *A Catalogue of French Medals in the British Museum*, ii, 1600–1672 (London, British Museum, 1988).

Michel Mollat du Jourdain, Monique de la Roncière, Marie-Madeleine Azard, Isabelle Raynaud-Nguyen and Marie-Antoinette Vannereau, *Les portulans: cartes marines du XIII au XVII siècle* (Fribourg, Office du Livre, 1984).

Roger J.P. Kain and Elizabeth Baigent, *The Cadastral Map in the Service of The State: A History of Property Mapping* (Chicago, University of Chicago Press, 1992).

Robert J. Karrow, Jr., *Mapmakers of the Sixteenth Century and their Maps: Bio-bibliographies of the cartographers of Abraham Ortelius* (Chicago, Speculum Orbis Press, 1992).

David Kingsley, *Printed Maps of Sussex, 1575–1900* (Lewes, Sussex Record Office, 1982).

Cornelis Koeman, *A World Map in Armenian printed in Amsterdam in 1695* in *Imago Mundi* 21 (1967), pp.113–15.

Cornelis Koeman and Marco van Egmond, 'Surveying and Official Mapping in the Low Countries, 1500–c.1670' in *The History of Cartography volume three. Cartography in the European Renaissance part two* (Chicago, University of Chicago Press, 2007), pp.1246–95.

Karl Kollnig, *Liselotte von der Pfalz Herzogin von Orléans, Eine fürstliche Münzsammlerin* (Melsunger, Gutenberg, 1987).

Joseph W. Konvitz, *Cartography in France 1660–1848: science, engineering and statecraft* (Chicago, University of Chicago Press, 1987).

Alex Krieger and David Cobb with Amy Turner (eds.), *Mapping Boston* (Cambridge, Massachusetts, MIT Press, 1999).

Karel Kuchař, *Early Maps of Bohemia, Moravia, and Silesia,* translated from the Czech original by Zdenek Safarik (Prague, 1961).

Helmut Kugler, Sonja Glauch and Antje Willing (eds.), *Ebstorf Weltkarte,* 2 vols. (Berlin, Academie Verlag, 2007).

Marcia Kupfer, 'Medieval World Maps: embedded images, interpretive frames' in *Word and Image* 10/3 (July–September 1994), pp.262–88.

Marcia Kupfer, 'The Lost Wheel Map of Ambrogio Lorenzetti' in *Art Bulletin,*78/2 (June 1996), pp.286–310.

Eugenio La Roca, 'The Newly Discovered Fresco from Trajan's Baths, Rome' in *Imago Mundi* 53 (2001), pp.68–79.

Toby Lester, *The Fourth Part of the World: the epic story of history's greatest map* (London, Profile, 2009).

J. Lieure, *Jacques Callot: Catalogue de l'œuvre gravé* (Paris, Editions de la Gazette des Beaux Arts, 1927).

G. van Loon, *Histoire Métallique des XVII provinces des Pays Bas depuis l'abdication de Charles Quint jusqu'a la Paix de Bade en MDCXIV,* 5 vols. (The Hague, 1732–7).

A. Macdonald, 'Plans of Dover in the Sixteenth Century' in *Archaelogia Cantiana* 49 (1937), pp.108–26.

Diary of Sir Frederic Madden (Bodleian Library, Oxford).

Jessica Maier, 'Mapping Past and Present, Leonardo Bufalini's Plan of Rome (1551)' in *Imago Mundi* 59:1 (2007), pp.1–23.

Claude-François Menestrier, *Histoire du Roy Louis le Grand Par les Medailles, Emblemes, Devises … Augmenté de 5. Planches* ('Paris, Nolin, 1691' [Amsterdam: anon.] 1691).

Peter H. Meurer, *Fontes Cartographici Orteliani: Das Theatrum Orbis Terrarum von Abraham Ortelius und seine Kartenquellen* (Weinheim, V.C.H. Verlagsgesellschaft, 1991).

Peter H. Meurer and Günter Schilder, 'Die Wandkarte des Türkenzuges 1529 von Johann Haselberg und Christoph Zell' in *Cartographica Helvetica* 39 (2009), pp.27–42.

Marica Milanesi, 'Antico e moderno nella cartografia umanistica: le grandi carte d'Italia nel Quattrocento' in *Geographia Antiqua* xvi/xvii (2007–8).

Ernesto Milano, commentary to *Planisfero Castiglioni: carta del navegare universalissima et diligentissima* (Modena, Bulino, 2002).

W. Minet, 'Some Unpublished Plans of Dover Harbour' in *Archaelogia* 62 (1922), pp.185–225.

Jan Mokre, 'Globen under freiem Himmel: Beispiele aus Wien' in *Der Globusfreund* 47/48 (1999/2000), pp.125–41.

Dominique Moran, 'Soviet cartography set in stone: the "Map of Industrialization"' in *Environment and Planning D: Society and Space* 24 (2006), pp.671–89.

Nigel J. Morgan, *Early Gothic Manuscripts II 1250–1285* [Survey of Manuscripts Illuminated in the British Isles IV] (London, Harvey Miller, 1988).

Juan del Campo Munoz, *Breve Historia del Palacio de Viso del Marques* (Madrid, Museo Naval, 1994).

Mauro Natale, *The Borromeo Islands and the Angera Fortress* (Milan, Silvana, 2000).

Kenneth Nebenzahl, *Maps of the Holy Land: Images of Terra Sancta through Two Millennia* (New York, Abbeville Press, 1986).

Howard Nelson and Yolande O'Donoghue, *Chinese & Japanese Maps* (exhibition catalogue: London, British Library, 1974).

K.S. Papazian, *Merchants from Ararat: a Brief Survey of Armenian Trade through the Ages* (New York, Ararat Press, 1979).

Mary S. Pedley, *The Commerce of Cartography: making and marketing maps in eighteenth-century France and England* (Chicago, University of Chicago Press, 2005).

Monique Pelletier, 'Des Globes pour le Roi-Soleil. Les origins des "globes de marly" in *Tours et Contours de la Terre. Itinéraires d'une femme au coeur de la cartographie* (Paris, Presses de l'école nationale des Ponts et Chaussées, 1999), pp.23–32.

Monique Pelletier, 'Le Cardinal, Le Moine, Le Roi et les autres. Les acteurs des globes du Roi-Soleil' in *Tours et Contours de la Terre. Itinéraires d'une femme au coeur de la cartographie* (Paris, Presses

de l'école nationale des Ponts et Chaussées, 1999), pp.33–45.

Monique Pelletier, *Cartographie de la France et du monde de la Renaissance au Siècle des lumières* (Paris, BNF, 2001).

Pliny, *Natural History* (Loeb edition. Cambridge, Harvard, 2003).

Wolfram Prinz, *Die Entstehung der Galerie in Frankreich und Italien* (Berlin, Gebr. Mann, 1977).

Jürgen Prüser, *Die Göhrde. Ein Beitrag zur Geschichte des Jagd- und Forstwesens in Niedersachsen* (Hildesheim, Lax, 1969).

Arthur H. Robinson, 'The President's Globe' in *Imago Mundi* 49 (1997), pp.143–52.

Margery Rose and Mary Ravenhill, *Devon maps and map-makers: manuscript maps before 1840* (Exeter, Devon and Cornwall Records Society, 2002).

Guy Roux and Muriel Lahaine, *Art et folie au Moyen Age: aventures et énigmes d'Opicius de Canistris (1296–vers 1351)* (Paris, Léopard d'Or, 1997).

C.R. Salisbury, 'An early Tudor map of the River Trent in Nottinghamshire' in *Transactions of the Thoroton Society of Nottinghamshire,* LXXXVII (1983), pp.54–9.

Heinz Sarkowski, 'Rudolf Kochs Deutschlandkarte als Politikum' in *Philobiblion* 43 (1999), pp.27–34.

Günter Schilder, *Monumenta Cartographica Neerlandica* II (Alphen-aan-den-Rijn, Canaletto, 1987).

Günter Schilder, *Monumenta Cartographica Neerlandica* VI (Alphen-aan-den-Rijn, Canaletto, 2000).

Günter Schilder, *Monumenta Cartographica Neerlandica* VII (Alphen-aan-den-Rijn, Canaletto, 2003).

Günter Schilder and Helen Wallis, 'Speed Military Maps Discovered' in *The Map Collector* 48 (1989), pp.22–6.

Richard Schofield, 'Old boundaries for a New State: The Creation of Iraq's Eastern Question' in *SAIS Review* XXVI (2006), pp.27–39.

Juergen Schulz, 'Jacopo de' Barbari's View of Venice: map making, city views and moralized geography before the year 1500' in *Art Bulletin* 60 (1978), pp.425–74.

Juergen Schulz, 'Maps as Metaphors: Mural Map Cycles of the Italian Renaissance' in David Woodward (ed.), *Art and Cartography. Six Historical Essays* (Chicago, University of Chicago Press, 1987), pp.97–122.

Juergen Schulz, *La cartografia tra scienza e arte. Carte e cartografi nel Rinascimento italiano* (revised edition of Italian translation originally published in 1990) (Modena, Panini, 2006).

Wilfried Seipel (ed.), *Der Kriegszug Kaiser Karls V gegen Tunis. Kartons und Tapisserien* (Vienna, Kunsthistorisches Museum, 2000).

Scott J. Seregny, *Russian Teachers and Peasant Revolution: The Politics of Education in 1905* (Bloomington and Indianapolis, Indiana University Press, 1989).

Rodney W. Shirley, *Early Printed maps of the British Isles, 1477–1650* (Castle Cary, Map Collector Publications, 1991).

Rodney W. Shirley, *The Mapping of the World: Early Printed World Maps 1472–1700* (Riverside, Early World Press Ltd, 2001).

Geoffrey Simcox, *Victor Amadeus II. Absolutism in the Savoyard State 1675–1730* (London, Thames & Hudson, 1983).

R.A. Skelton, 'A Contract for World Maps at Barcelona, 1399–1400' in *Imago Mundi* 20 (1968), pp.107–113.

R.A. Skelton, *County Atlases of the British Isles 1579–1850. A Bibliography* (Folkestone, Dawson, 1970).

R.A. Skelton, *Maps: A Historical Survey of Their Study and Collecting* (Chicago, University of Chicago Press, 1971).

Thomas R. Smith, 'Manuscript and Printed Sea Charts in Seventeenth Century London: The Case of the Thames School' in Norman Thrower (ed.), *The Compleat Plattmaker: Essays on Chart, Map and Globe Making in England in the Seventeenth and Eighteenth Centuries* (Berkeley, University of California Press, 1978), pp. 45–100.

Martin P. Snyder, *City of Independence: views of Philadelphia before 1800* (New York, Praeger, 1975).

David Starkey (ed.), *The Inventory of King Henry VIII. i, The Transcript*, (London, Society of Antiquaries, 1998).

Francis Steer, *A Catalogue of Sussex Estate and Tithe Award Maps* (Sussex Record Society vol. Lxi) (Lewes, Sussex Record Society, 1961).

Richard Talbert (ed.), *Ancient Perspectives: Maps and Their Place in Mesopotamia, Egypt, Greece, and Rome* [16th Kenneth Nebenzahl Jr. lecture in the History of Cartography] (Chicago, University of Chicago Press, forthcoming).

Dan Terkla, 'The Original Placement of the Hereford Mappa Mundi' in *Imago Mundi* 56 (2004), pp.131–51.

Rodney Thomson, 'Medieval Maps at Merton College Oxford' in *Imago Mundi* 61 (2009), pp.84–90.

Simon Thurley, *Whitehall Palace. An architectural history of the royal apartments, 1240–1690* (New Haven and London, Yale University Press, 1999).

George Tolias, *The Greek Portolan Charts 15th–17th Centuries* (Athens, Olkos, 1999).

Sarah Toulouse, '*L'Hydrographie Normande*' in Monique Pelletier (ed.), *Les Couleurs de la Terre* (Paris, Seuil/BNF, 1998), pp.52–5

Sarah Toulouse, 'Marine Cartography and Navigation in Renaissance France' in David Woodward (ed.), *The History of Cartography volume three. Cartography in the European Renaissance part two* (Chicago, University of Chicago Press, 2007), pp.1550–68.

Hilary L. Turner, 'A wittie devise': the Sheldon tapestry maps belonging to the Bodleian Library, Oxford' in *Bodleian Library Record* 17 no. 5 (April 2002), pp.293–313.

Hilary L. Turner, 'The Sheldon Tapestry Maps: their Content and Context' in *The Cartographic Journal* 40/1 (June 2003), pp.39–49.

Hilary L. Turner, 'Oxfordshire in Wool and Silk: the tapestry map of Oxfordshire woven for Ralph Sheldon "the Great"' in *Oxoniensia*, lxxi (2007), pp.67–72.

Hilary L. Turner, 'Tapestries once at Chastleton House and their Influence on the Image of the Tapestries Called Sheldon: A Reassessment' in *Antiquaries Journal* 88 (2008), pp.313–46.

Sarah Tyacke, *London Map-Sellers: A collection of advertisements for maps placed in the* London Gazette *1668–1719* (Tring, Map Collector Publications, 1978).

Vladimiro Valerio, *Società Uomini e Istituzioni Cartografiche nel Mezzogiorno d'Italia* (Florence, Istituto Geografico Militare, 1993).

Victoria County History of England. Sussex vi, T.P. Hudson (ed.) (London, University of London, 1986).

Dirk de Vries, 'Dutch Cartography' in Robert P. Maccubbin and Martha Hamilton-Phillips (eds), *The Age of William III & Mary II. Power, Politics and Patronage 1688–1702* (Williamsburg, College of William and Mary, 1989), pp.105–11.

Helen M. Wallis, 'The First English Globe: A Recent Discovery' in *The Geographical Journal* cxvii (1951), pp.275–90.

Helen M. Wallis, 'Further Light on the Molyneux Globes' in *The Geographical Journal* cxxi (1955), pp.304–11.

Helen M. Wallis, 'Globes in England up to 1660' in *The Geographical Magazine* xxxv (1962), pp.267–79.

Franz Wawrik and Elisabeth Zeilinger (eds), *Austria Picta: Österreich auf alten Karten und Ansichten*

(exhibition catalogue: Vienna, Austrian National Library, 1989).

James Welsh, *Military Reminiscences* (London, 1830), p.243.

Peter J. Whitehead, 'The Marcgraf Map of Brazil' in *The Map Collector* 40 (1982), pp.17–20.

Peter J. Whitehead and M. Boeseman, *A Portrait of Dutch 17th Century Brazil: Animals, plants and people by the artists of Johan Maurits of Nassau* (Amsterdam, Oxford, New York, North-Holland Publishing Company, 1989).

Simon Winchester, *The Map that Changed the World: the tale of William Smith and the birth of a science* (London, Viking, 2001).
Jürgen Wilke, *Die Ebstorf Weltkarte*, 2 vols. (Bielefeld, Verlag für Regionalgeschichte, 2001).

Armin Wolf, 'News on the Ebstorf World Map: date, origin, authority' in Monique Pelletier (ed.), *Géographie du Monde au Moyen Age et à la Renaissance* (Paris, Comité des Travaux Historiques et Scientifiques, 1989), pp.51–68.

David Woodward, *Maps as Prints in the Italian Renaissance: makers, distributors & consumers* [The Panizzi Lectures 1995] (London, British Library, 1996).

Lawrence Wroth, *Abel Buell of Connecticut* (Middletown, Wesleyan University Press, 1958).

Lord Howard of Effingham and the Spanish Armada. With exact facsimiles of the 'Tables of Augustine Ryther AD 1590 [i.e. the plates in 'A Discourse concerning the Spanishe fleete inuadinge Englande' by P. Ubaldini, engraved by A. Ryther after Robert Adams], and 'The Engravings of the Hangings of the House of Lords' by John Pine [after Cornelius Vroom], AD 1739. With an introduction by Henry Yates Thompson (London, Roxburghe Club, 1919).

Cordell D.K. Yee, 'Traditional Chinese Cartography and the Myth of Westernization' in J.B. Harley and David Woodward (eds), *The History of Cartography volume two. Cartography in the traditional East and Southeast Asian societies book two* (Chicago, University of Chicago Press, 1994), pp.170–202.

Kees Zandvliet, 'Art and Cartography in the VOC Governor's House in Taiwan' in Paula van Gest-van het Schip and Peter van der Krogt (eds), *Mappae Antiquae Liber Amicorum Günter Schilder* (Amsterdam, Hes & De Graaf, 2007), pp.579–94.

Simone Zurawski, 'New sources for Jacques Callot's *Map of the Siege of Breda*' in *The Art Bulletin* lxx (1988), pp.621–39.

INDEX

Note: illustration references marked in **bold**

Académie des Sciences 146, 152
Adams, John 118
Adams, Robert 101
Adele of Blois 14
Afghanistan 116, 162
Africa 25, 52, 79, 89, 161
Agas, Ralph 121
Agnese, Battista 34
Agrippa, Marcus Vipsanius 12, 160
Akrotiri 11
Albemarle, George Keppel, 3rd Earl of 110
Alberti, Leon Battista, *Trattato di Architettura* 27
Alexandria 13
Alfonso V, King of Portugal 52
Alford, John 141
All-Russia Exhibition of Industry and Art 156
Almagià, Roberto 10
Alpers, Svetlana 10, 96
Alsace 163
America 84
America, Spanish 74
America, United States of 64, **64**, 112, 142
Americas, the 60
Ammaedara 13
Amman 13
Amsterdam 50, 68, 122, 124
Amyce, Israel 118
Anne, Queen of England 27, 44, 141
Ansbach 108
Antoniszoon, Cornelis 40
Antwerp 40, 50, 89, 120
Apian, Peter, *Liber Cosmographicus* **15**, 16
Apian, Philip 24, 64, 158; *Chorographia Bavariae* **23**
Armazém de Guinéa, the 34
Armenia 124, 164
Ascona 14
Ashmolean Museum 77
Ashridge 79
Asia 25, 52, 84
atlas: of Europe, 98–9, **98–9**; Klencke, 92, **92–3**
audience chambers, as settings for maps 48–76
Augsburg 16, 21, 25, 48, 50, 54–5, **54–5**, 71
Augustus, Emperor 12, 44, 160–1
Austria 159
Austrian National Library 13
Averham 106, 107

Babylon 13
Bacon, G.W. 162
Baltic Sea 128
Bandenburg 65
Barbaro family 160
Barcelona 78
Bavaria 24, 64, 158
bedchambers, as settings for maps 76–81
Beijing 94
Belgrade 32
Bellini, Leonardo 52
Belvedere Villa 26, 77
Bender, Richard *see* Koch, Rudolf
Berlin 92
Bern 160
Besson, Jean **35**
Biblioteca Marciana, Venice 48, 52
Bibliothèque nationale de France 62
Bignon, Jean Paul 152
Birkholz, Daniel 76
Blackness 110

Blaeu, Joan 42, 94; *Brasilia qua parte paret Belgis* **42–3**
Blaeu family 92
Blaeu, Willem 152
Blasco, Michelangelo 74–5; *Stato di Milano e sue provincie tirate dal Censimento o sia perticato di esso stato...* **74–5**
Blenheim Palace 24
Bocquet, Francois-Nicolas 152
Bodleian Library 106
Bogislav XIV, Duke of Pomerania 65
Bohemia 70–3
Bologna 25
Bombay 9, 96
Bonne, Rigobert 89
Bonsignori, Stefano 24, 38, **39**, 82
Booth, Charles, *Descriptive Map of London Poverty* 134
Borgonio, Giovanni Tommaso 35
Boston 112–13
Boteler family 148
Boyne, the, Battle of 44
Brandenburg, Frederic Wilhelm I, Elector of 92
Bratislava (Pressburg) 32
Braun, Georg 27
Brazil **42–3**, 74
Breda 24, 68; Siege of **68–9**
Bridger, Elizabeth 141
Bridger, Richard 140–1
Bristol 118
British Library: map of Scotland in 9; Henry VIII's map of Italy transferred to 28; Cottonian map in 50, 103; Agas survey maps in 121; Janssonius Europe map in 128
British Museum 85, 142
British Ordnance Survey 74, 130
Brueghel, Pieter 138
Bruges 21, 124
Brunswick, Dukes of 80
Buckingham Palace 21
Budgen, Richard 130; *An Actual Survey of the County of Sussex* **130–1**
Buell, Abel 64; *A New and Correct Map of the United States of North America* **64**
Buen Retiro Palace 68
Bufalini, Leonardo, 86, *Roma* **86–7**
Bunker Hill, Battle of 112
Burano 48
Burgerzaal, Amsterdam 50
Burghley, William Cecil, Lord 19, 50, 107
Burt, William 160
Buti, Leonardo 24, 38
Buytewech, Willem 118
Byzantine empire 13

Cabinet War Rooms 101
cabinets, as settings for maps 82–99
Cabot, John 144
Callot, Jacques 24, 68, 120; *The Siege of Breda* **69**, 120
Camden, William 56
Camocio, Gioan Francesco 40
Canada 84
Cantelupe, St Thomas, Bishop of Hereford 14
Canterbury 140
Canton 145
Capitoline Museum 11
Caprarola 20
Carlos II, King of Spain 122
Carthage 13
Cartier, Jacques 84
Cartographia Bavariae **108**
Cary, J. *Cary's Pocket Globe* **95**
Castiglione, Baldassare 16; *Il Cortegiano* 120
Cavendish, Thomas 60
Cecil, Sir William *see* Burghley, William Cecil, Lord

Ceylon 166
Chaplin, Charlie **18**
Charles, Duke of Lorraine 24
Charles de Bourbon, Cardinal 24
Charles II, King of England 42, 92, 96, 103, 122, 132, 142, 146
Charles V, Holy Roman Emperor 24, 25, 36, 54, 71, 120
Charles VI, Emperor 150, 160
Chatsworth 19
Cheney, Lord Thomas 121
China 60, 94, 124, 166
Chirikov, Igor Ivanovich, *Carte topographique de la frontière turco-persane...* **116–17**
Christina, Queen of Sweden 96
Churchill, Sir Winston 19, 165
Citolino, Alessandro 92
Civil War, English 56
Claesz, Cornelis, and Johannes Janssonius, *Europa* **128–9**
Claudius Ptolemy *see* Ptolemy, Claudius 16
coins 96–7
Coke, Viscount 32
Colbert, Jean-Baptiste 132
Colle Oppio, Rome 12
College of Bonhommes 79
Cologne 50, 78
Colonial Office, the 101
Columbus, Ferdinand 30
Comestor, Peter, *Historia Scholastica* 79
Committee of Plantations, the 101
Constantinople (Istanbul) 13, 24, 26, 32, 164
Cook, James 152
Cordell, Sir William 118
Cork 103
Coronelli, Vincenzo 160
Cortesi, Paolo 50
Cosimo I de Medici *see* de Medici, Cosimo I, Grand Duke of Tuscany
Cotton, Sir Richard *see* Rogers, John
Cotton, Sir Robert 28
Cottonian map 50, 102; *see also* Griffith, Maurice
Court of Augmentations 79
Covens and Mortier (map publishers) 120, 150
Cowdray Park 77
Cranach, Lucas, the Elder 126
Crete 11
Crimean War 116
Cruchley, George Frederick, 149; *Cruchley's enlarged Map of Europe* **149**
Culloden 110
Cumberland, The Prince William Augustus, Duke of 110, 150
Cunningham, William 55
Cuzco 89
Czech Republic 71
Czechoslovakia 159

Daily Courant (newspaper) 44
Dandolo, Francesco, Doge of Venice 48
d'Annebault, Claude 84
Danti, Egnazio 24, 82, 98
Darby, John 138; *Plan of the parish of Smallburgh* **138–9**
Daumont, Jean Francois 152
de Barbari, Jacopo 16, 21, 25, 30, 54, 118; Venice MD **30**
de Bazan, Alvaro, Marques of Santa Cruz 27
de Bello, Richard, Prebendary of Haldingham and Sleaford 14
de Guevara, Alonso Ladrón 68
de Hooch, Pieter, 118; *A Woman Drinking with Two Men* **118**
de Hooghe, Romeyne 146
de Jode, Gerard 89; *Nova Totius Terrarum Orbis Juxta Neotericorum Traditiones* **88–9**

de Laralde, Andries 120
de Medici, Cosimo I, Grand Duke of Tuscany 23, 24, 82
de Medici, Cosimo, the Elder 82
de Medici, Francesco, Grand Duke of Tuscany 27, 38
de Montmorency, Anne 84
De origine urbium Italie 28
de Pannemaker, Willem **22**, 24
de Pisan, Christine 76, **76**
de Saint-Marceauxin, René 160
de Vuauconsains, Anthonie 90
Denmark 163
Desceliers, Pierre 21; *see also* world maps
d'Estrées, César, Cardinal 160
Devonshire, Peregrine Cavendish, 12th Duke of 19
Dias, Manuel 94; and Nicolo Longobardi, *Chinese Terrestrial Globe* **94**
Dieppe 84, 85
Disraeli, Benjamin, Earl of Beaconsfield 162, 164
Doge's Palace, Venice 20, 27, 28, 48, 49, 101
Don Basin: Heart of Russia, The, 166, **167**
Dover 104, 105
Dr Strangelove (film) 19
Drake, Sir Francis 60, 96
Drogheda 44
Dron, Henri, *L'Europe des Points Noirs* **163**
Drouais, Francois-Hubert, *The Comte de Vaudreuil* **102**
Dumbarton 110
Dürer, Albrecht 30

East India Company, British 9, 52, 101, 114, 145, 166
East India Company, Dutch 50, 101, 124
East Indies 60
Ebstorf world map *see* world maps
Eckhout, Albert 42
Edinburgh 110
Edmund, Earl of Cornwall 79, 80
Edward I, King of England 77, 79, 80
Edward VI, King of England 79
Elizabeth I, Queen of England 19, 56, 60, 101, 146
Elliot, Sir Henry 116
Elphinstone, John 110; 'A New Map of North Britain...' **110**, 111
Elstrak, Remond 25
Elyot, Sir Thomas 16; *Boke named the Governour* 105
Emanuele Filiberto, Duke of Savoy *see* Savoy, Emanuele Filiberto, Duke of
England 24, 25, 50, 60, 62–3, **62–3**, 142, 144
Englisch, Brigitte 13
Este family 16, 24, 50
Euclid, *Elements* 16
Eusebius, Bishop of Caesaria 13
Europe 122, 124, 149, 164
Evelyn, John 90, 92, 118
Evesham world map *see* world maps

Farnese, Cardinal Alessandro 20–1, 23, 49–50
Farnese, Palazzo **9**; *see also* Sala del Mappamondo
FDR *see* German Federal Republic
Felbrigg Hall, Norfolk 121
Ferdinando, Grand Duke of Tuscany 24
Ferrara 16, 24
Field of the Cloth of Gold 24
Finland 164
Fiorani, Francesca 10, 25
First World War 116, 164
Fitzhugh, Augustine 144; *Chart of Newfoundland and the Grand Banks* **144**
Florence 21, 26, 27, 68; topographical map of 38–9
Folkestone 105

Fontainebleau 24, 46; *Galeries des Cerfs* 22, 24
Forma Urbis Romae **11**
Fort, James 103
Fra Mauro World Map *see* world maps
France 24, 103, 122, 124, 144, 159, 163
Francis I, Holy Roman Emperor 74
Francis I, King of France 77
Frankfurt 50
Frazer, William 52
Frederick II, King of Prussia 150
Frederick William, Elector of Brandenburg 42
Freud, Lucian 19
Frisius, Simon 36, **36–7**
Frobisher, Sir Martin 60
Fugger family 54

Galerie des Glaces, Versailles 27
Galleria delle Carte Geografiche see Vatican
Garda, Lake 101
Gardner, William 130
Garret, John 82, 118
Gastaldi, Jacopo 19, 40
Genoa 26
George, Prince of Denmark 27
George I, King of England 46, 91
George II, King of England 46, 150
George III, King of England 21, 46, 103, 130, 145
Georgia 64
Germain, Jean **14**
German Federal Republic (FDR) 146
Germaniae (L'Empire D'Allemagne) **151**
Germany 71, 80, 122, 142, 150, 158, 163–5
Gertruydenberg 68
Gervase of Tilbury 80
Gill, Eric 166
Gill, MacDonald 166; *Tea Revives the World* **166**; *Wonderground Map of London* 134, **137**
Gladstone, William 162
globes 60–1, 94–5, 160
Gloucestershire 56
Godfather III (film) 19
Goehrde, the, map of 46–7
Goethe, Johann Wolfgang von 158
Gonzaga family 16, 20, 50
Gough Map 106
Great Britain 40, 74, 116, 164
Great Dictator, The (film) **18**, 19
Great Exhibition (1851) 160
Great Globe, Tilly Whim **160**
Great Yarmouth **50–1**
Greene, Robert 142, 146
Gregory XII, Pope 25
Griffith, Maurice, Archdeacon of Rochester 102; 'Cottonian' map of the British Isles **102**
Grym, Sigmund 54–5
Guangzhou 145
Guardaroba Nuova, Palazzo Vecchio 82–3, **82**
Guildhall 132
Guinness Book of Records 92

Habsburg Empire 70
Habsburg family 27
Haine & Son (publishers) 146
Hakluyt, Richard, *Divers voyages touching the discouerie of the Americas* 20
Halliwell, A.E., 166; *All British Routes*, 166, **167**
Ham House **82**, 98
Hampton Court Palace 22, 24–5, 50, 77, 91, 141
Handel, George Frideric 91
Hanover 76, 91
Hanseatic League 128
Harvey, P. D. A. 77

Haselberg, Johann, 32, **32**
Hatfield House 21
Henri II, King of France 24, 46, 84
Henri IV, King of France 24, 46, 90
Henry III, King of England 48, 76–80
Henry V, King of England 148
Henry VII, King of England 28
Henry VIII, King of England 21, 22, 24–5, 28, 30, 40, 50, 77, 82, 102
Hermitage Museum **17**, 19
Herodotus 78
Herrenhausen 91
Hertz, Johann Daniel 70
Higden, Ranulf, *Polychronicon* 148
Hill, Francis 140; *A Map And description of all ye Lands belonging to Richd Bridger...*, **140–1**
Histoire de l'Entree de la Reyne Mere du Roy Tres-Chrestien dans la Grande Bretaigne **77**
Hitler, Adolf 159
Hogarth, William 110
Hogenberg, Franz 27
Hohenzollern dynasty 108
Hölderlein, Friedrich 158
Holland 142
Holme, Thomas 142; *A Map of the Settled Part of Pennsylvania*, 142, **142–3**
Holy Land 13, 126, **127**, 146, 164; *see also* Palestine
Homem, Diogo 34, 40
Homem, Lopo 34
Hondius, Jodocus 61, 94
Hooke, Robert 132
houses, domestic, as settings for maps 118–145
Howard, Charles, Lord High Admiral, Lord Howard of Effingham 101
Hugh of St Victor 79
Hughli River 114
Hugo, Hermannus 68
Hungary 71
hunting forests, mapped 46

Ignatiev, Nikolai Pavlovich, Count 116
Imperial Library (Austrian National Library) 160
India, British map 9
Innocent VIII, Pope 26
International Postal Union 160, **161**
International Tea Market Expansion Board 166
Ipswich 138
Iran 124
Ireland , 44–5, 103
Isabella, Infanta 68
Isabelle, Queen of France 76, **76**
Isidore, Bishop of Seville 13, 28, 78
Italie Provincie Modernus Situs **29**
Italy 24, 28–9, 71, **74–5**, 82, 122, 161, 163

James I, King of England 62
James II, King of England 44, 103
James VI, King of Scotland 21
Janson, C., *Plan du Chateau et Jardin Royal à Herrenhausen...* **91**
Janssonius, Johannes 37; *see also* Claesz, Cornelis
Jefferys, Thomas 98
Jerome 13
Jerusalem 13, 24
Joao III, King of Portugal 25
Johanna, Archduchess 27
John, Duke of Marlborough 24
Jordan 13
Jukes, Francis 112

Karachi **147**
Kauffer, Michael 71
Kelham 106

Kensington Gardens 91
Kensington Palace 27, 44, 141
King, Martin Luther 160
Kinsale 103
Kinsale Harbour, Map of **103**
Klencke Atlas *see* Atlas
Klencke, Johann 42, 92
Koch, Rudolf 158–9; and Fritz Kredel, Richard Bender and Berthold Wolpe, *Deutschland und angrenzende Gebiete* **158–9**
Kolkata 114
Kredel, Fritz 159; *see also* Koch, Rudolf

La Rochelle, Siege of 120
Laicksteen, Peter 126; and Christian Sgrooten and Hieronymus Cock, *Nova Descriptio Amplissimae Sanctae Terrae* **127**
Lake Worth 160
Lamb, Francis 142
Lapérouse, Comte de, Jean-François de Galaup 18
Lastman, Pieter 146
Lea, Philip 118
Leemput, Remigius 77
Leibniz, Gottfried Wilhelm 91
Leicester, Robert Dudley, 1st Earl of 56
Leonardi, Antonio 28, 49, 50
Leopold I, Holy Roman Emperor 35
Levasseur, Guillaume 85
Leybourne, William, *The Compleat Surveyor* 121
Lintott, Bernard 130
Lintott, Henry 130
Lisbon 25
Livy 12
Lombardy 28, 74
London 50, 78, 118, 132, 134–6, 160
London Gazette (newspaper) 44
Long Compton 56
Longobardi, Nicolo 94; *see also* Dias, Manuel
Lorenzetti, Ambrogio 48, 101
Lorraine, Charles, Duke of 68
Louis XIII, King of France 128
Louis XIV, King of France 27, 35, 96, 122, 132, 160
Louis XVI, King of France 18
Louvre Museum 83
Lovat, Simon Fraser, Lord 110
Lubin, Erhard 64–5; *Nova illustrissimi Principatus Pomeraniae Descriptio* **65–7**
Lüneburg 80
Luvinius, Martinus 88
Lyon 78

Macchiavelli, Niccolò 16
Madaba 13
Madrid 24, 68
Magellan, Ferdinand 23
Maine 64
Major, R.H. 85
Mantua 16
mappae mundi 14, **14**, 17, 48, 76–7; *see also* world maps
Mappa Mundi, Hereford 14, 76, 78, 79, 80
Marcgraf, Georg 42; *Brasilia qua paret Belgis* **42–3**
Marcus Vipsanius Agrippa *see* Agrippa, Marcus Vipsanius
Maria Theresa, Archduchess of Austria, Holy Roman Empress 74
Marinoni, Johann Jakob 74
Marlborough, John Churchill, 1st Duke of 103
Martini, Simone 49
Mary I, Queen of England 34
Maurits, Johan, Count of Nassau 42, 92
Mauro, Fra 48, 52
Max Emanuel, Elector of Bavaria 24
Maximilian I, Holy Roman Emperor 30

medals 96–7
Medici family 50
Mediterranean Sea **22**, **34**, 128
Melford Hall, Suffolk 118
Mercator, Gerard 89, 96
Mercator, Michael 96
Mercers' Hall 132
Merchant Adventurers 128
Meurer, Peter 10
Mexico City 89
Michaelangelo Buonarotti 152
Milan 16, 74–5, 101
Milanesi, Marica 10
Millo, Antonio 40; *Tuto el discoperto in carta marina in piano* **40–1**
Mills on the River Trent Nottinghamshire **106–7**
Milos 40
Minoans, as mapmakers 11
Minoritas, Fra Paolino 28
Mirabeau, Honoré-Gabriel Riqueti, Comte de 165
Mississippi River 64
Modena 24
Molenaer, Jan Miense 124; *The Artist's Studio* **120**; *The Schoolroom* **146**
Moll, Herman, *A Correct Globe with ye Trade Winds* **95**
Molyneux, Emery 60–1
Monsiau, Nicolas-André 18; *Louis XVI giving instructions to La Perouse, 1785* 15
Moor, Dimitri, *Be On Your Guard* 166, **167**; *The Hunger Spider* 166, **167**
Morgan, William 134; *London* **132–3**
Morley, Lord 138
Morosini, Andrea 48
Morris, William 159
Mortlake 56
Moscow 156
Mosley, Sir Oswald 159
Moxon, J., *Pocket Globe* **95**
Müller, Johann Christoph, 70–1; *Mappa geographica regni Bohemiae in duodecim circulos divisae...* **70–3**
Murray, Lord George 110
Muskham 106
Mussolini, Benito 159, 161

Nancy 68
Naples 26, 74, 101
Napoleon III, Emperor of France 163
Nassau, Johan Maurits, Duke of *see* Maurits, Johan
National Maritime Museum 21
Negri, Professor Cristoforo 85
'Nemesis', *The overthrow of His Imperial Majesty King Jingo I* **162**
Nero, Emperor 12
Netherlands, the 21, 25, 68
Neumont, Maurice, *La Guerre est l'Industrie Nationale de la Pruses* **165**
New England 64
Newark 106–7
Newfoundland 144
Newton and Berry's New Terrestrial Globe **95**
Niccoli, Niccolò 82
Nicholas II, Tsar of Russia 156
Nolin, Jean-Baptiste 152; *Le Globe Terrestre* **152–5**
Normandy 25
Norwich 55, 138
Nöttelein, Georg 108; *Große Wald und Fraischkarte von Nürnberg* **109**
Nottingham 107
Nuremberg 108
Nuti, Lucia 10

Ogilby, John 118, 132

open air, as setting for maps 160–7
Orange 11, 12
Orange Cadastre **11**, 12
Ormonde, James Butler, 2nd Duke of 44
Ormonde, James Butler, Earl of 103
Orosius 13
Ortelius, Abraham 88–9; *Nova Totius Terrarum Orbis Juxta Neotericorum Traditiones* **88–9**; *Parergon* 89
Otto the Child, Duke of Brunswick 76
Otto IV, Holy Roman Emperor 76
Ottoman Empire 32, 116
Oxfordshire, Sheldon map of, 56, **56–9**

Pakistan 116
Palazzo Pubblico, Siena 20, 48–9, **49**
Palazzo Vecchio, Florence 27, 38, 82, 98; *Sala delle Mappe Geografiche* 23
Palazzo Venezia, Rome 49
Palestine, mosaic map of **13**
Panizzi, Sir Antonio 85
Paris 24, 50, 77, 78, 90
Paris, Matthew 104
Paris, Treaty of 144
Parker, Sir Philip 138
Parnell, Charles Stuart 162
Pasi, Marco Antonio 24
Paul II, Pope 49
Pavlovitch, General Nikolai 116
Payne, William 102
Pelham, Sir Henry 130; *A Plan of Boston in New England with its Environs* 12, **113**
Pelham, Peter 112
Penn, William 141–2
Pennsylvania 141, 142
Pepys, Samuel 118
Perry, Grayson 19; *Map of Nowhere*, 81, **81**
Persia 116, 124, 164
Peru 85
Pétain, Marshal Henri-Philippe 165
Peter the Great, Tsar of Russia 156
Petty, Sir William 44
Peutinger, Konrad 54
Peutinger Table, the 13, 54
Philadelphia 64
Philip the Good, Duke of Burgundy **14**
Philip II, King of Spain 19, 36, 40, 126
Philip II, Duke of Pomerania 65
Philip IV, King of Spain 68
Pierce, Mark 118
Pine, John 101
Pinturrichio (Bernardino di Betto di Biago) 26, 82
Pisa 26
Pliny the Elder 78; *Historia Naturalis* 12–13, 16
Plummer, Samuel 130
Poland 159, 164
Polo, Marco 23, 52, 89
Pomerania 64, 65–7
Pompeii 12
Pomponius Mela 14, 16, 28
Portraiture of Philadelphia 142
Portugal 34, 144
Post, Frans 42
Pourbus, Pieter 21
Prado Museum 24
Pratt, Henry 44; *Tabula Hiberniae Novissima et Emendatissima* **44–5**
Pressburg *see* Bratislava
Prussia 163, 165
'Psalter' world map *see* world maps
Ptolemy, Claudius 16, 52, 89; *Geographia* 12, 14, 16, 28, 82
Punjab **147**

Raleigh, Sir Walter 60
reception rooms, private, as settings for maps 76–81

Reiner, Wenzel Lorenz 70
Rembrandt van Rijn 124, 146; *Self-portrait* **121**
Rémy de la Fosse, Louis 46
René, Duke of Lorraine 25
Reynolds, Charles 9
Rhineland, the 163
Richard, King of the Romans 80
Ringcurran Castle 103
Roberval, Jean-François de La Rocque, Sieur de 85
Robinson, William 103
Rodriguez, Luigi 101
Rogers, John, and Sir Richard Cotton, *Ffor dovour pere* **105**
Roman Empire, 13, **161**
Romans, as mapmakers 11
Rome 11–13, **13**, 20, 26, 30, 86–7, 163
Roosevelt, Franklin D. 19
Rose, Fred W. 162, 164; *Comic Map of the British Isles* **162**; *Serio-Comic War Map* **164**
Rosselli, Francesco 21, 26, 38
Rostock 92
Rotterdam 128
Roy, General William 9
Royal Academy of Arts 112
Royal Exchange 132
Royal Palace, Amsterdam, *Burgerzaal* **20**, 21
Royal Palace, Madrid 24
Royal Society 132
Rubens, Peter Paul 68
Rudolf II, Emperor 94
Russia 116, 128, 156, 163, 164, 166
Russian Empire 147
Ryther, Augustine 101

St Petersburg **17**
Sala Bologna, Vatican 77, 82
Sala del Mappamondo **9**, 23; Farnese Palace 49–50; Palazzo Pubblico **49**, 101; Palazzo Vecchio 82
Sala delle Carte Geografiche, Uffizi Palace 38
Sala dello Scudo, Venice **48**
Salzburg, Archbishop's Palace in 10, **10**
Sandby, Paul 9, 102, 112
Sandby, Thomas 110
Sanderson, Sir William 60
Santa Maria del Giglio 160, **160**
Santorini 11
Sardinia 12
Savoy, Emanuele Filiberto, Duke of 25
Savoy, Estates of **35**
Savoy, Victor Amadeus II, Duke of 35
Saxe-Weimar, Duke of 94
Saxton, Christopher 24, 25, 50, 121, 138
Schenk, Petrus 122
Schilder, Günter 10
Schleswig 163
schoolrooms, as settings for maps 146–159
Schoonebeck, Adrian 124; and Peter Schoonebeck, *Double Hemisphere World Map in Armenian* **124–5**
Schoonebeck, Peter *see* Schoonebeck, Adrian
Schulz, Juergen 10
Schutz, Augustus 150
Scotland 9, 25
Seckford, Thomas 138
Secretary of State's Room, as setting for maps 101–117
Seld, Jörg 16, 21, 25, 55; *Sacri Romani Imperii Civitas Augusta Vindelicorum* **54–5**
Seville 36–7, 128
Sgrooten, Christian, 126; *see also* Laickstein, Peter
Sheldon family 118
Sheldon, Ralph 56

Sheridan, Thomas, *School for Scandal* 98
Sicily 74
Siena 38, 48, 101
Singleton Copley, John, 112; *Boy with Squirrel* 112
Sistine Chapel 152
Sloane, Sir Hans 142, 144
Smallburgh 138
Smith, John 'Warwick', *Dover* **104**
Smith, William 147
Society of Arts 130
Society for the Diffusion of Useful Knowledge 147
Society for the Propagation of Christian Knowledge 147
Sophia, Electress of Hanover 91
Sorte, Cristoforo 101
Spain 71, 82, 103, 144
Spanish Succession, War of 35, 96
Speed, John 25, 62, 146; *Map of England, Wales and Ireland* **62–3**
Spinola, General Ambrogio 68
S.P.K., *Confiance – ses amputations se poursuivent méthodiquement* **165**
Stalin, Josef 19
Stanza di Raffaello, Vatican 152
Stará Boleslav 71
Stigliola, Nicola Antonio 101
Stirling 110
Strabo 14, 16
Stuart, Prince Charles Edward 110
Stuart dynasty 62
Stuart, Prince James 110
Suleiman I, Grand Sultan (Suleiman the Magnificent) 32
Sussex 130
Sutton, Sir William (?) 106–7
Swanage, Dorset 160
Sweden 65
Switzerland 14

Taiwan 50
Talbert, Professor Richard 13
Tangiers 96
Tapuya, the 42
Tavoletta Pretoriana 74
Temple of Peace 11
Terrazza delle Matematiche, Uffizi Palace 24, 27; *see also* *Sala delle Carte Geografiche*, Uffizi Palace
Terza Loggia, Vatican 10, 23, **26**, 27, 82
Thera 11
Thornton, John 142, 144
Tiberius Sempronius Gracchus, General 12
Toddington 121
Tomasik, M.I., *Pictorial Map of European Russia* **156–7**
Toskanatrakt **10**
Trent, River 106–7
Trotsky, Leon 166
Tudor dynasty 62
Tunisia 13
Tupinambu 42
Turin 35
Turkey 164

Uffizi Gallery 83
Uffizi Palace 24, 27, 38
USSR (Union of Soviet Socialist Republics) 19
Utrecht 68
Utrecht, Treaty of 144

Valk, Gerard 122; *Nova Totius Terrarum Orbis Tabular/ Europa/ Asia/ Africa/ America* **122–3**
Vanandetsi, Archbishop 124
van Baerle, Caspar, *Rerum per Octennium in Brasilia... Historia* 42
van den Wijngaerde, Anthonis 36, 77

van der Goes, Hendrik 120
van Hoogstraten, Samuel, *View of a Corridor* **119**
van Langeren, Florent 61, 94
van Orley, Bernard **24**, 25
Vasari, Giorgio 82
Vatican 10; *Galleria delle Carte Geografiche* **20**, 23–4, 82; maps of Bologna in 25
Vavassore, Giovanni Andrea 120
Velázquez, Diego 24, 68
Veneto, the 28
Venice 16, 21–8, 30–1, 40, 48, 54, 101, 118, 160
Vermeer, Jan 118
Vermeyen, Jan Cornelisz 24
Verona 101
Verrazzano, Giovanni 40
Verrazzano, Girolamo 40
Versailles, Palace of 27, 160
Vespasian, Emperor 11
Vesuvius 12
Victoria, Queen of England 162
Vienna 70, 74, 160; Siege of 32–3
View of Guangzhou (Canton), A **145**
Vincent, William, Dean of Westminster 52
Visconti family 16
Viso del Marqués 27, **27**
Von Czoernig, Carl 147
Vyne, the, Hampshire 118

Waldseemüller, Martin 25
Wales 24, 25, 50, 62, **62–3**
Walter, Stephen 19, 134, 136; *The Island* **134–5**
Warminghurst, Sussex 140
Warwickshire 56
Wasserkirche, Zurich **83**
Weiditz, Hans 16, 21, 25, 55; *Sacri Romani Imperii Civitas Augusta Vindelicorum* 54
Welser family 54
Welsh, James 9
Wenceslas, King of Bohemia 71
Westminster 76
Westminster Palace 79, 80, 101
Weston, Warwickshire 118
Whampoa 145
Whitehall Palace 20, 22, 25, 28, 62, 92, 94, 132
Wilhelm I, Kaiser 164
Willdey, George 98, 118
William I, King of England 14
William III, King of England (William of Orange) 44
Winchester Palace 48
Winchester, Simon 147
Wirsung, Markus 54, 55
Wolf, Armin 76
Wolfart, Hans 65
Wolgast, Philip Julius of 65
Wolpe, Berthold 159; *see also* Koch, Rudolf
Wood, Mark, 114; *Survey of the country on the eastern bank of the Hugly, ...* **114–5**
Woodward, David 120
Worcestershire 56
world maps 40–1; Ebstorf map 76, 78, 80–1, **80**; Evesham world map 148, **148**; Descelier world map, 84–5, **84–5**; Duchy of Cornwall map 78–80, **79**; Fra Mauro World Map **52–3**; 'Psalter' world map 76, 78–80, **78**; *see also* mappae mundi
World War I *see* First World War
Woutneel, Hans 62
Wyld, James 160

Yeakell, Thomas 130

Zell, Christoph **32**
Ziarnko, Jan, *Ville Citte Universite de Paris* **90**
Zorzi, Alessandro 49
Zurich 50